How to Breathe Underwater

HOW TO BREATHE UNDERWATER

FIELD REPORTS FROM AN AGE OF RADICAL CHANGE

Chris Turner

BIBLIOASIS

WINDSOR, ONTARIO

FIRST EDITION

Library and Archives Canada Cataloguing in Publication

Turner, Chris, 1973-, author
    How to breathe underwater / Chris Turner.

Issued in print and electronic formats.
ISBN 978-1-927428-75-7 (pbk.).--ISBN 978-1-927428-76-4 (ebook)

    1. Human ecology. 2. Sustainability. I. Title.

GF50.T87 2014   304.2   C2014-903804-6
                                              C2014-903805-4

Edited by Jeet Heer
Copy-edited by Allana Amlin
Typeset by Chris Andrechek
Cover designed by Kate Hargreaves

Biblioasis acknowledges the ongoing financial support of the Government of Canada through the Canada Council for the Arts, Canadian Heritage, the Canada Book Fund; and the Government of Ontario through the Ontario Arts Council.

PRINTED AND BOUND IN CANADA

Contents

For Ashley, who taught me how to travel—and why

INTRODUCTION
THE RADICAL CHANGE BEAT

I GUESS THIS BEGAN with a stack of *Rolling Stone* magazines in the attic of my aunt's house in Toronto, unearthed during a summer visit. Almost literally so—the magazines encrusted with other piles of old reading clutter, some of them wilted and browned by several muggy summers. They seemed, to my adolescent eyes, like artifacts of some sort. It was probably 1985 or 1986 and I was maybe thirteen. It was an idle curiosity that became a passion and then a craft and finally a career.

There were dozens of the magazines, hundreds maybe. Years and years of them in wonky leaning-tower piles. I remember a few grand old decadent Grateful Dead and Rolling Stones '70s images, but mostly there were grinning covers starring 1980s icons—Tom Cruise and Michael J. Fox and Bruce Springsteen. I knew what *Rolling Stone* was, though to that point the only music magazines I'd ever bought for myself were metalhead titles like *Hit Parader* and *Circus*. I read record reviews at first, skimmed celebrity profiles. The tone, more than anything else, was what drew me deeper into the pages. Here was a world so much more sophisticated and deeper in texture than the one I knew as a military brat enduring adolescence in remote northern towns. Here, despite Cruise's megawatt cover-boy grin, was a celebration of originality and defiance and oddness as core virtues, an offhand eggheadedness about pop music and blockbuster movies, a critical language with which begin to imagine an adult life deeply unlike my current predicament.

That fall, I traded my superannuated *Sports Illustrated* subscription for the cover of the *Rolling Stone* in my mailbox every other week. I soon started to recognize bylines, names I knew were significant but had no idea why. Tom Wolfe, Hunter S. Thompson, Greil Marcus, Annie Liebovitz. I still read the record reviews at the back of the book and the gossipy "Random Notes" full of boldface names at the front, but each issue drew me deeper into the part of the magazine I would one day understand is referred to by magazine pros as "the well." Big, meaty, multi-thousand-word features. Vivid scenes from the other side of the world, anguished front-line environmental reports, political screeds full of ideas I recognized were important but—as with the bylines—did not yet quite get why.

I remember in particular a story by Randall Sullivan about the murder of a California cheerleader by her ostracized classmate. It seemed to say something vital and unequivocal about the age I was living in—and felt like I was drowning in, though I could not find a way to articulate an SOS. I didn't know I cared about California or cheerleaders or why they hated and hurt each other, but the story pulled me in and made me care more about these cheerleaders in this particular California suburb than I ever had about anything in the current-affairs world of the daily news. This, as I would come to understand intimately, is the transcendent power of a great magazine feature: it can turn a single specific sliver of time and place into the axis of the whole world.

My favourite *Rolling Stone* writer at the time—here I date myself even more hopelessly—was P.J. O'Rourke, who in those days was still a sort of gonzo wise-ass foreign correspondent and not the clumsy parody of a braying La-Z-Boy Baby Boomer he's become. In my undergraduate years, I found better music and smarter friends, and I came to know the wider horizon of top-tier narrative non-fiction. I read *Spin* because it understood the 1990s alternative culture I was wading into by then so much better than the aging *Stone* did. I picked up *Details* primarily for the long, sharply written dispatches from post-Communist Eastern Europe and the U.S. election trail and Hollywood's adult film community. These were inevitably written under the byline that would, above any other anywhere, inform my own work in the years to come: Chris Heath. My favourite history professor pointed me in the direction of *Harper's*, an ancient title that was entering a resurgent golden age as a platform for distinctive non-fiction voices. It was on those pages that I found the byline that became the impossible literary benchmark which I would, like many young writers of my generation, envy and ape in equal measure and never come close to equalling: David Foster Wallace. And at some point, by some chance or happenstance I can't recall, I started religiously reading the magazine I most wanted to write for, which folded before I could beg my way into a job. It was a strange beast called *Might*, founded by another writer whose work I would come to know intimately and emulate often: Dave Eggers. It was wise and sincere and satirical and mocking all at the same time, which approximated perfect pitch for the mid-1990s.

Meanwhile, I scorned the newsy squares at my school newspaper, wrote occasional screeds on the state of pop culture for other campus publications, worked with a few friends to co-publish two issues of a photocopied, stapled, pre-internet zine. (Hey, it was The Nineties.) I graduated knowing I wanted to write for a living but having no clue how to do so. I decided to enroll in journalism school, and chose Ryerson University in Toronto because I figured the city, headquarters of every Canadian magazine I'd ever heard of, was the best place in the country to be as a young magazine writer. This was perhaps the only sober and wholly sound judgement I made in all of 1996.

At Ryerson, I learned at least as much about the business of journalism as I did about how to craft a story, which is the inverse of a knock on the program. Journalism is a job whose skills you acquire through experience, repetition, churning out copy and paying your dues. What I needed was some sense of the dimensions of a newsroom, the cadence of a production meeting, the structure of a story pitch. And that's what I got. I applied for an internship at the *Globe and Mail* after my first year and was deemed worthy of an interview, but I somehow managed not to notice I'd been so chosen until after it was scheduled to happen (see the note above about sound and sober judgement in those years). Even though the *Globe* graciously rescheduled, I didn't get the job (with good cause) and I was already choosing another kind of career path. Thorny and twisted. Here be serpents. *Freelance.*

At some point in that first hazy year at Ryerson, I'd discovered a Canadian magazine with some of the energy and cultural assuredness

I'd been hunting for since my first chance encounter with *Rolling Stone*. It was a magazine that talked a lot about computers and the internet, which I didn't really care much about, and about Marshall McLuhan, who I knew I was supposed to care more about that I did. But more than that it had real voices in it. It was fully immersed in its subject. It clearly had ambitions of being what any cohort other than the jaded, suspicious one coming of age in the 1990s would've called the voice of its generation. The magazine was called *Shift*. Miraculously, a classmate in the magazine stream at Ryerson was just finishing up an internship there. (Here I'll start giving full credit wherever it's due—thanks, Felix Vikhman.) He was working part-time, for actual (though meagre) pay. The magazine had wildly ambitious young founders—full credit and then some to Evan Solomon and Andrew Heintzman—and some serious new investors, as well as the hottest design team in the business (more credit: Carmen Djunko and Malcolm Brown).

I was on time for *that* internship interview. I bragged about reading *Harper's* cover to cover, which was nearly true, even though I'm sure the senior editor who eventually brought me into the *Shift* fold—the excellent, exacting Joanna Pachner—didn't buy it. Anyway, I got the gig. I started working there in the spring of 1998, even before my final classes at Ryerson concluded. I skipped graduation. I had my diploma, in the form of a *Shift* stipend that didn't even cover all my rent. I felt like I more than deserved the position—my ego had really bloomed at Ryerson—but at the same time I couldn't believe my luck. I still can't.

LET ME SET THE SCENE HERE. I'm enough of a product of my wise-ass generation to reflexively retch at the mere sight of Boomeresque sepia, but this is a scene worth setting nevertheless. The Balfour Building on Spadina, just south of Queen, is a grand old Garment District pile of bricks, half-renovated in those days but still cheap enough for a shoestring-budget magazine. Exotic "new media" companies that seemed to change name and ownership every other month had offices on a couple of other floors. Sony Music had a marketing office or something on another floor for a time. There were a few great warehouse bars on and just off King Street farther west. The Amsterdam

Brewery's take-home store was a five-minute speedwalk away, and a key intern's task on warm, sunny days was to stake out a good table or two on the Black Bull patio. *Shift's* office was on the second floor, consisting mostly of a single vast warehouse space with a "board-room" you built by closing off a section of the floor using pull-down garage doors. The digital team—one of the first dedicated groups in any magazine office in Canada, headed by the gifted duo of Barnaby Marshall and Dave Sylvestre—occupied the psychogeographic centre of the room, flanked by the designers and photo department on one side and the non-executive editors and us interns and editorial assistants on the other. The digital team had also claimed command of the officewide soundtrack—I recall a lot of Portishead and Radiohead and the Propellerheads, the heady symmetry of which only strikes me now. Everyone's email inbox had the same notification sound on it—a simple, old-school ping—and you knew a great conversation or raging debate was underway when the "reply all" pings came undulating across the space and back again in rapid succession, like the sudden eruption of a flock of agitated birds. Production meetings often involved uproarious arguments about what the magazine was, about what "digital culture" was, about what it meant to be alive and awake at this exact moment.

I know how that sounds now. I *know*. But dammit, it was true. We thought we just might conquer the whole world. *We* would be the next *Wired*, the new *Rolling Stone*. Our ambitions for ourselves and the magazine were boundless.

In the meantime, we subsisted on cheap Vietnamese subs and dive coffee shop fare and the increasingly elaborate hors d'oeuvres at the more and more lavish parties Toronto's newborn dotcoms were throwing. As the only magazine fully on the digital beat, we were invited to everything. A company called Digital Renaissance—later known as ExtendMedia—was handed a pile of Bell R&D money at some point and threw a party in 1998 out at its vast warehouse space in what's now known as Liberty Village that was full-on fall-of-Roman in its decadence. There was a single-malt scotch bar. An *open* single-malt scotch bar.

It was an odd-angled life of decadent poverty to be at *Shift* in those years. Hunter S. Thompson, patron saint of freelance magazine writing, once summed up the early years of his career like this: "Yeah, drunk, horny and broke. Somehow there were 48 hours in a day and 18 days in a week. But the suffering of going through ten years of it. 'Freelance journalism'—that sounds romantic now, right? But the desperation—teetering from one word to another." It was like that. I remember giddiness and frustration, abundance and deep poverty, all at once and in equal measure. Only in retrospect do I see how singularly, momentously, stop-me-before-I-go-all-*Summer-of-69* wonderful it all was.

I thought every office I would ever work in from that first day forward would be like *Shift*'s office in the Balfour Building in the strange, sainted summer of 1998. Little did I know that no other office would come close. I used to elbow Clive Thompson and Douglas Bell—both senior editors, both fine mentors, both accomplished writers in their own right—out of the way to check email. My fellow interns included Sheila Heti (who everyone knew even then was bound for the literary greatness she's achieved) and Ian Connacher (still my closest ex-*Shift* compadre, director of the trailblazing documentary *Addicted to Plastic*). Kevin Siu started in accounting; he's now a deputy editor at the *Globe*. Rolf Dinsdale was our ad sales director; he missed becoming the Honourable Member for Brandon-Souris by a couple hundred votes last year. Carmen Djunko and Barnaby Marshall are the reason you've heard of the Drake Hotel. Laas Turnbull and Liane George, who not long ago took over Toronto's *Eye Weekly* and turned it into *The Grid*, the best Alt-Weekly 2.0 in North America, first worked together in that office. (*I thought it would last forever / Those were the best days of my life.* I know, I know, I *know*.)

As GREAT AS THE PLACE WAS, the Balfour Building incarnation of *Shift* was as transient as the dotcom boom it rode to local/regional/very-nearly-international fame. The magazine's parent company went bankrupt in 2000, and *Shift* acquired stodgier ownership and less dynamic offices and finally ceased publishing in 2003. The legacy for me was the beat it put me onto, the one I followed down one meandering path or another to produce the essays and reportage in this volume. At *Shift*, pretty much by accident, I became a documentarian of radical change, of cultural upheaval and technological explosion and sudden, dramatic reversals and inversions of fortune. Two editors deserve credit for taking the gambles that brought me fully onto that beat: Laas Turnbull and Neil Morton.

Turnbull called me into his office shortly after he'd taken over as *Shift*'s editor-in-chief from Evan Solomon, who was off to TV stardom. This was the fall of 1998, and I was an editorial assistant—a full-time employee for part-time pay, making rent by transcribing interviews for Doug Bell's book (Toronto Book Award finalist *Run Over*) and the odd meaningless freelance gig. (Those celebrity-worship magazines they give away free at movie theatres? Yeah, that kind of thing.) Turnbull asked me what I really wanted to do. I said my student loans wanted a steady editorial gig but my gut wanted badly to write. He encouraged me to go with my gut, which is to say he dismissed me from my editorial responsibilities, let me continue to squat at *Shift*'s office during working hours, and soon handed me my first assignment. A Toronto software company called CryptoLogic was actually making money on the internet—all but unheard of in those days of venture capital angels betting on distant-future profits. I was sent to a bland office block far uptown to check it out. I came back and told Laas the real story was in the Caribbean, where the internet gambling boom fuelled by CryptoLogic's gaming software had triggered a gold rush on some tiny island called Antigua. My first feature assignment, I argued, would be to go to a tropical island paradise. I'd have been insulted at the time if Turnbull hadn't bought it, but in retrospect it seems miraculous he took the chance. That story, "Flipflops, a Desktop and One Billion Reasons Never to Leave," struck gold at that year's National Magazine Awards. Ready or not, my freelance career had launched.

Fast-forward a couple of years. Neil Morton was now at the helm of *Shift*, and I'd just finished a series of feature reports for *Time* magazine on the revolutionary impact of digital technology on Canadian society. The work at *Time* had often been frustrating—it was very much an editor-driven (and editor-written) magazine, not a writer's one—and the subject matter felt inconsequential. The dotcom bubble had popped, fortunes made and lost, and, yes, we all had communication and connectivity and self-publishing tools at our fingertips that had been unimaginable just a few years earlier. But still, the nagging question: What did it all matter?

I remember driving up El Camino Real in Silicon Valley, off from one lightweight interview to another as I rounded up profiles of Canada's most influential techies. I got stuck in a traffic jam at one point, inching along in lurches and halts, gazing out the window at strip-malled office-parked sprawl that could've been Anywhere, U.S.A. What did it all *matter*, if it was all just the same-old-same-old plus gadgets? I'd sat the day before with Jeff Skoll, whose time was not easy to procure and who would soon go on to a brilliant philanthropic second career and establish the production company responsible for *An Inconvenient Truth* and *Syriana* and the film version of *Fast Food Nation*. And we'd talked about all he'd done to make it easier to buy and sell garage-sale bric-a-brac on eBay.

Back in Toronto at a Firkin pub on Bay Street, I spat out all my self-loathing and misgivings to Morton over a couple of pints. There had to be a way to tell the story of technology no one was telling, the one about how we'd as yet done next to nothing as a civilization to confront the existential challenge of climate change. The rant was rambling and tangential. Somehow Morton found the will to assign me the task of turning it into an essay. I think we agreed on 5,000 words or so; I turned in somewhere near 10,000, writing in fuming and anguished and desperate late-night sessions. (My wife to this day refers to it as my "nervous breakdown piece.") It was published under the title "Why Technology Is Failing Us (and How We Can Fix It)" in September 2001. It won two National Magazine Awards, including the President's Medal for General Excellence, which in those days was given to the feature of the year. (It now goes to Magazine of the Year.)

I'd pivoted at some fundamental level in that essay, away from digital gadgetry and toward the infrastructure of a green economy, though it'd be a few more years before the rant became a beat and I became a full-time sustainability reporter. In the interim, Morton assigned me another sprawling five-figure-word-count dot-connecting essay—the simple task of summing up "ten years in the life of the culture" for *Shift*'s tenth anniversary. A crazy, exhilarating, terrifying blank cheque of an assignment. Reckoning there was a *Simpsons* quote for every occasion, I used lines from the show as the framing device. The resulting essay, "The Simpsons Generation," landed me an agent and worldwide book deals; *Planet Simpson* was published in 2004. As soon as it was done, I started back on the climate beat. I decided to chase solutions instead of causes and consequences, and uncovered the first young shoots of the green economy that would provide material for my next two books, *The Geography of Hope* (2007) and *The Leap* (2011). As I mentioned, *Shift* folded in the interim, but fortunately an upstart magazine emerged to fill the void in Canadian general interest titles—I believe the initial model was to be a Canadian *Harper's*—and I found my second feature-writing homebase on the pages of *The Walrus*.

NOT LONG AGO, I wrote a 3,000-word piece on the legacy of Oliver North's contra war in Central America for *Hazlitt*, a web magazine overseen by Random House Canada. The story would've been regarded, in the early days of *Shift*, as a "short feature." When *Hazlitt's* Twitter feed announced its publication, it referred to the piece as a *#longread*. Such is the precarious state of narrative non-fiction in the digital age that it must be flagged, a warning issued. This will not be digested quickly. Its purpose is not to convey a soundbite or a handful of data points. This isn't news. Its primary intent is *literary*, not informative.

I became a magazine writer because of *#longreads*. Every writer of literary non-fiction you've ever heard of is known to you because of *#longreads* so lengthy there's almost nowhere left in magazine publishing that would run them. If 3,000 words warrants the hashtag, we need a whole other name for the soaring literary form I've attempted to emulate in these pages. Jon Hersey's "Hiroshima" took up the entire August 31, 1946, issue of *The New Yorker*. Tom Wolfe and Norman Mailer and Joan Didion made their reputations on stories with word counts on the far side of 10,000. Some of them consisted of multiple parts, sprawling across several issues. Hunter S. Thompson built his rockstar-sized fame on an epic two-part stream-of-consciousness retelling of a series of debauched weekends in Las Vegas for *Rolling Stone*. Malcolm Gladwell, Eric Schlosser, Susan Orlean, Elizabeth Kolbert, Rebecca Solnit, on and on—all writers whose foundational work (and often their best) was not the book but the very long magazine piece the book was based on.

The form is as distinct as a novella, often as far from a book as a short story is from a novel. It captures a moment, reveals a cross-section, takes a snapshot portrait of a time and place. In my experience, great magazine pieces typically need 5,000 words minimum to hit their stride, and often only achieve a fully immersive narrative by a couple thousand words past that. The pinnacle of the form verges on novella-length, roughly 10,000 to 30,000 words. Gay Talese's 1966 *Esquire* masterpiece "Frank Sinatra Has a Cold," which many in the business (though not I) deem the best magazine piece ever, runs to 15,000. Tom Wolfe's "Kandy-Kolored Tangerine Flake Streamline Baby," considered the progenitor of the whole "new journalism" age of longform magazine writing, is more than 12,000 words. And "Post-Orbital Remorse," the report that led to Wolfe's book *The Right Stuff,* was longer still, stretching across four issues of *Rolling Stone.* Hunter S. Thompson took two issues, each installment clocking in on the far side of 20,000 words, to unspool the piece I actually consider possibly the best magazine feature ever: "Fear & Loathing in Las Vegas." It vies for the title with Joan Didion's "The White Album" (12,000-plus words) and David Foster Wallace's "Shipping Out" (20,000 words or so). In the latter piece, the footnotes alone would amount to a *#longread* by contemporary digital standards.

I'm belabouring the point about word counts because pretty much all the stories in this collection came in longer than assigned and ran far longer than many magazines will even abide nowadays. Because we're in danger of losing a whole art form by attrition and commercial myopia. Because on dark days I feel like I spent 15 years honing an obsolete craft. And because all of these worries are compounded for the Canadian writer.

The limits on Canadian magazine writing are as old as the country itself. They are products of geography, money and a lingering colonial dependency. It's expensive to send magazines to a small population scattered across such a massive nation, expensive as well to send writers out to report in its farflung precincts. There have never been a lot of venues for literary non-fiction in this country, and there have perhaps not been this few since the early days of continent-wide rail. So much easier to rely on the vast American market or the better-funded British one to feed the ex-colony its true stories.

Not long ago, Noah Richler wrote in the *National Post* about a conversation he had with Ian Jack, the former editor of the venerable

British literary journal *Granta*. The magazine was known for doing omnibus nationalistic overviews of literary and essay-writing giants. Was it Canada's turn? Richler wondered. "Jack was all for it," he wrote, "but evidently pained. He didn't think there were enough good Canadian essayists around."

Well, as a two-time winner of the National Magazine Award for Best Essay, let me respond in my best colonial English: That's just bullshit, Jack. To his credit, Richler goes on to list a wide range of top-tier Canadian journalists and essayists, Douglas Coupland and Charlotte Gray and John Vaillant and Margaret MacMillan among them, who are every bit the equal of any other nation's literary cohort. I could add any number of bylines just in the pages of the magazines from which these pieces were drawn—Clive Thompson, Lynn Cunningham, John Lorinc, Andrea Curtis, Rachel Giese, Curtis Gillespie, Charlotte Gill—but that would miss the point. Canada has plenty of great non-fiction writers; it has precious few non-fiction *venues*, nowhere near enough places to hone the craft and build a readership and reputation. Those who can manage the jump (from Noah's father Mordecai to the aforementioned bestselling Gladwell to current *Esquire* star writer Chris Jones) ply their trade in the United States or farther afield. And in the meantime Canadians, I fear, just might find cause to buy into Ian Jack's flawed line of reasoning, thinking we lack the cultural wherewithal to produce great essayists.

Venues matter, and markets matter. With too few venues, it's hard to keep a career afloat on feature-length narrative non-fiction. You graduate to books (even if sometimes your premise can't sustain a book's length) or you churn out riffs and rants instead of fully exploring a line of thinking at essay length. With such small markets, even those magazines that can run ambitious features can't run very many of them, and the limited and shrinking ad revenues slash budgets and page counts and oblige editors to shoehorn great 7,500-word drafts into 4,000-word holes in their wells. When yet another American non-fiction writer comes in for lavish praise—one of the current critical darlings is John Jeremiah Sullivan—I don't find myself envious of his literary mettle so much as his ready access to word counts. That's not to say Sullivan hasn't earned his praise—he has—but name a Canadian magazine with the resources to send a writer to Jamaica for the sole purpose of attempting, possibly in vain, to track down Bunny Wailer for an interview. Here's my tally based on current market conditions: _____. Sullivan's resulting piece ("The Last Wailer," 8,400 words in the January 2011 issue of *GQ*) is one of those great immersive surprising magazine features, the kind that lured me into the business in the first place. And I count myself blessed that I've found a couple of Canadian outlets capable, at least intermittently, of providing the travel budget and white space to write a few of my own. But I do find myself wondering more and more about all the ideas that didn't fit, the pieces that didn't get written and the long features that got hacked away and became shadows of their best selves because sometimes for years on end there was nothing in this country resembling a genuine mass-market general interest magazine.

To be sure, the radical change beat sometimes offered up just the right mix of newsworthiness, character and place to tidily fit into one of the country's narrow feature story slots. The stories collected here are the ones that *didn't* get away. A digital casino boom on a Caribbean island with fewer inhabitants than Barrie, Ontario; an entire city built from scratch to create a second Silicon Valley in the Malaysian wilderness; a remote First Nations community on the pristine British Columbia coast caught in the perilous trajectory of oil tanker traffic. These are the sorts of stories where you can bank on the "there" being there before you pack your bags and start racking up expenses. The pieces in this collection I consider the most successful, though, were stitched together from scraps and stolen moments while reporting on less momentous stories. *Time* magazine's tech-mongering gave me enough background detail to rant about its limits and blindspots. *Shift* magazine's tenth anniversary provided the outsized space to talk about the big currents reshaping society. A lecture tour of Australia offered a handful of passing glimpses of the whole fragile planet's radically altered climate.

I became adept, I think by necessity, at tying together disparate phenomena and distilling common themes from mismatched pots. There was so rarely any time to spend absorbing a single time and place. And even when I did find the time, there was often no way to tie those scenes neatly into one of the larger themes. Radical change is amorphous, multivalent, its trajectory clear only in retrospect. In the midst of the churn, there's often not enough to hold on to and piece together into the sort of story pitch that earns one of those precious slots in Canada's few remaining general-interest feature wells.

Looking back, I find myself obsessed most of all with all the stories I never told, all the detail that never quite fit. A representative example: In Malaysia in late 1999, on the hunt for the heart of the nation's grandiose "Multimedia Super Corridor," I spent my free time at a Kuala Lumpur internet café, answering email and reading through background research. The clientele was young, almost exclusively male, and evenly split between young men playing *FIFA 2000* and other young men cruising chatrooms for short-term male companions from abroad. The café's proprietor explained about Malaysia's strict social taboo on homosexuality, the brief escapes from it that could be won by hooking up with a visiting foreigner and racing off to some beach resort that catered to Western social mores. This was a scene as ripe with meaning as anything I found in Cyberjaya, the country's half-built tech industry hub, but it didn't quite fit into the assignment.

The incidental detail was sometimes the most telling, but it often fell outside the conventional story arc of a feature with a tight word count for a publication with a precise mandate.

On assignment for *Time* in Silicon Valley, say, listening as a Cisco executive fed business-as-usual platitudes into my recorder from a boardroom looking out over a half-empty post-bust cubicle farm.

That itching sense of a scene too forced as I drank "Warp Core Breach" cocktails with a handful of code-writing acquaintances at the Trekkie-themed bar of the Las Vegas Hilton between sessions at a hacker conference.

This game we *Shift* interns would play at the peak of Toronto's dotcom boom, seeing who could get the first straight answer about what the company plying us with drinks and sushi at their launch party actually made.

Watching the first dawn of 2000 from a houseboat bobbing between islands on Ha Long Bay, Vietnam, chuckling with friends over the failure of the Y2K virus to do anything significant to the magnificent scenery.

The freak snowstorm in the Himalayan foothills a few months later, and the way every local was certain this was something that'd literally *never* happened before in their lifetimes.

The things the climate scientists told you once the recorder had been turned off that made you fear for your children's very lives as sure as if they were toddling too close to a cliff's edge.

One story of this age of radical change is a collection of these vertiginous moments of confusion and dissonance, transient scenes caught midway between euphoria and despair and unsure which way the balance would tip, certain only that the centre would not be holding much longer. That was a story I never found a way to pitch properly. I'm still not sure if it's a hopeful story or a dystopic one, or if more precisely there's such a thing as an optimistic dystopia. Maybe you can find it in the pauses and line breaks in the stories that follow.

THERE IS NO STORY I FAILED to tell more fully in these years than the one about the year I spent living in India from the summer of 1999 to the summer of 2000, watching the world's second most populous nation embody the tectonic shifts in communications, commerce and climate that animate the rest of the stories in this collection. I never told it because I never quite grasped it with any certainty. India was, in a sense, the subject that proves the lie of every magazine piece ever written, that whopper about how the storyteller actually has a complete and nuanced understanding of his subject, or about how any subject worth talking about can be reduced to 5,000 words. Or really about how there's such a thing in messy racing reality as a single theme or a tidy narrative arc.

I never wrote about India directly because the only thing I knew for certain was that I didn't really understand what was happening there or what any of it meant. I could describe what I saw, but I couldn't figure out how it fit into a coherent narrative. I saw a billion arcs. I didn't know which one was the true story. So I told none of them.

Here's a true story about India at the dawn of the age of radical change that seems, fifteen years on, like the truest arc I saw—the untold story alluded to by each of the ones in this collection. It starts in New Delhi, in a neighbourhood called Paharganj—a grimy warren of market stalls and cheap hotels across from the city's main railway station. Paharganj, like many backpacker ghettos, is India in microcosm and yet not really India at all. Or both somehow. But unlike some of the world's backpacker ghettoes, India will not consent to being cordoned off into autonomous zones. Even its backpacker ghetto is profoundly Indian. Metaphorically speaking, no density of party-friendly hostels and German-themed bakeries in Paharganj will be sufficient to stop a sacred cow from taking a shit in your path.

Anyway, my wife and I were living for the year up in the old Raj-era summer capital of Shimla in the Himalayan foothills, but assorted travels and arriving and departing visitors and business of the visa renewal sort took us often to Delhi. We tried a wide range of hotels all over the city, from the emerging middle-class enclaves of South Delhi to enticing but impossible-to-find oases deep in the heart of the old city near the Jama Masjid. The mix of convenience and location kept us coming back to Paharganj often as not, and we eventually settled on a tidy, welcoming little spot called the Royal Guest House as our default home base. Part of the reason we liked the Royal was because it was just a few doors down from the Hotel Gold Regency, which was just beyond our meagre means as a place to stay but had the neighbourhood's best internet café.

We came to know the Gold Regency in discrete episodes, split apart by weeks or months like time-lapse photographs. The first encounter was in the fall of 1999. The Gold Regency's management had set up a couple of computers at kiosks in the lobby for its predominantly foreign clientele to check email and make travel plans. They were always in use; often we would have a bite to eat in the hotel cafeteria waiting for one to come free. The joint turned out a passable macaroni and cheese that was like heaven to dairy-nurtured palates that had tasted only paneer and yogurt for months on end. (The short list of our deepest cravings at the end of a year on an Indian diet read: *Cheese / good booze / salsa / bacon / more cheese.*)

Months later: The Gold Regency lobby had been almost entirely converted into an internet café that spilled over into the cafeteria itself, desktop terminals filling every spare square of tile throughout the hotel's ground floor. Receipts from the cafeteria now identified it as the "Cyber-Bar-B-Q," though I never heard anyone call it that out loud. The clientele had begun to shift as well, the terminals being used by a roughly equal mix of backpackers (seemingly every such internet-seeking customer in all of Paharganj) and young Delhi residents. I once caught a glimpse on a neighbouring monitor of a website bedecked with glitzy party trim and blinking text—the first online arranged-marriage service I'd ever seen, but certainly not the last. An eager mother and bored-looking daughter were seated together in front of it in dressy *salwar kameez*. You'd also sometimes see young men staring hungrily at porn, hard as you might try to avoid it. And despite the dozens of new terminals row on row, the place was packed often enough to perpetuate an ill-advised love affair with passable mac-and-cheese.

Months later: To most visitors, the Hotel Gold Regency must surely by now have seemed like an expansive multi-level internet café that also rented out a few rooms. The rows of computer kiosks wandered up staircases, across landings, down a broad corridor that had once been maybe a banquet hall or meeting room. Only the mac-and-cheese, still solidly passable, remained unchanged.

Here was the unpredictable, explosive growth of digital communication in microcosm, the unforeseen interactions between cultures and classes happening in what the hardcore digerati called "meatspace" as well as online. Here was a proverbial genie, the bottle forever shattered, spreading like an amorphous sci-fi blob into every nook and cranny. A whole great subcontinent still largely paced to pre-industrial rhythms lay not far in the past, could still be found a short rickshaw ride away. Telephones, to the majority of India's people, remained specialized devices that resided in public callboxes or office spaces on retail streets, for use on serious business and special occasions only. A future of unprecedented growth and software engineering ingenuity and half the world's call centres and back-office tech support services (not to mention the world's fastest-growing mobile phone market) lay not far ahead.

The India we'd come to know in the waning months of 1999 was already a place from another age. And the fulcrum of this radical change was the internet café at the Hotel Gold Regency in Paharganj. I should've pitched some magazine on 24 hours in the life of the place, gone full *#longread* on it like Neal Stephenson did with his magisterial *Wired* feature about the laying of the world's fibre optic cables (another in my personal Top Ten best magazine stories ever).

Instead, I checked my email and ate passable mac-and-cheese. I never wrote about it until now. Perhaps it's the unseen fulcrum for this collection of stories as well. Every story is a compromise between intent and execution, between what is observed and what is reported, between the tidy verbatim of what a subject says and the complex web of what she seems to mean. Between an entire boisterous moment in a time and place and the snapshot that fits into the diminishing word count of a *#longread*.

The stories might be compromised, the high word counts harder by the issue to maintain, the whole romantic desperation of the free-lance game—to mash up a phrase from the good Dr. Thompson's summary—growing less romantic and more desperate with each month's mortgage payment. But still the form draws me back time and again. It is an art all its own, and I've delighted in honing my craft on the journeys and flights of intellectual discovery contained in this collection and many others besides. And I'm still most satisfied, as a writer, with those rare, giddy moments when the fates align and the magazine's available space is sufficient and the subject worthy enough, all of it combining by some lucky alchemy into a finished story of singular and self-contained beauty. Few books (and none of my own) seem so true and consistent in tone, so concise and vivid in detail, so artfully *sturdy*. Literary non-fiction, the true-life novella, a great *#longread*, a magazine feature—a solid, handcrafted work of literary art, a stolen snapshot of a time and place that might, if only for a moment, seem like the fulcrum around which all the world is turning.

I

DISPATCHES FROM THE DOTCOM FRONTIER

My older brother was the techie. He's the reason I knew my way around a Commodore 64, learned the few simple lines of code that would fill a DOS screen with endlessly repeating lines of text. I never thought much of it all as a culture until I arrived at *Shift* magazine in Toronto to start my internship in 1998. It probably didn't help that I came through the doors of the Balfour Building down on Spadina for the first time with a big ole chip of generic Gen-X cynicism on my shoulder. I was predisposed against buying any kind of hype. I'm not sure if that made me a better reporter on the dotcom beat, but in any case it eventually taught me humility, because I'd colossally misjudged the power of ones and zeroes to reorder pretty much everything.

Shift was way ahead of the digital curve, which meant that by a process of accelerated osmosis I got most of the way there myself. It was my job at one point to try to cajole everyone from Neil Postman and Terry Gilliam to Nicholas Negroponte and Bruce Sterling into participating in the magazine's "State of the Net" forum. I investigated usenet forums and fan fiction, studied the internet's first attempts at online travel brokerage and dot-commerce, received a cease-and-desist letter from the Church of Scientology while documenting its failed attempt to prevent anti-Scientology information from spreading online. I was *Shift*'s first video game columnist—and thus one of the first in the country. I once knew an awful lot about how to earn make-believe money playing Ultima Online and then sell it for hard currency on eBay. Doing copy-edits on a short profile of the founder of a newfangled "weblog" called Metafilter, I discovered the website that would become my digital water cooler for the next decade (until Twitter came along).

As digital technology spread from margin to mainstream, I got my first big assignments beyond *Shift*'s pages writing about similar phenomena for business magazines and national newspapers and even *Time* magazine. I interviewed legendary entrepreneurs and engineers and scientists (Jeff Gosling, Jeff Skoll, John Polanyi, the founders of Rockstar Games), profiled the progenitors of passing trends (the creator of Doodie.com, the fleeting titans of Ottawa's dotcom bubble), and fabricated an interview with the star of the Pac-Man video game reboot (I decided he was wistful and ornery, like Danny Bonnaduce in his reality TV phase). I like to think I developed pretty strong skills at intervening in obscure subcultures and digesting technical detail and—hardest of all—teasing lively quotes out of engineers, all of it on the fly.

The most rewarding work in those years, three examples of which follow, was the fully immersive stuff, those out-of-body experiences born of parachuting into another socioeconomic universe and trying to absorb it as fully as possible during a brief visit and then describe the sensation with some verisimilitude afterward. In retrospect it's an obvious lesson, but I remember finding it curious at the time how the best tech stories came not from sitting in front of a screen exploring cyberspace but from going to the physical spaces where the digital world was being built. Even in the internet age, being there mattered.

FLIPFLOPS, A DESKTOP AND ONE BILLION REASONS NEVER TO LEAVE

Shift, May 1999

I was the envy of every underemployed twentysomething I knew—jetting off to a tropical paradise to write about gambling as my first serious journalism assignment. The reality of Antigua beyond the resort's gates and the tight itineraries of the cruise ships was less grand—the capital, St. John's, was a gritty, isolated, underdeveloped town. When I got off my flight from Toronto, I was the only one who didn't board a package-resort-bound shuttle bus. There was no guide, no itinerary, no safety net.

As with many life-changing experiences—sex, scuba diving, dodgy home-made wine—you never forget your first time. I had no idea what I was doing. I was terrified, beyond clueless. The Shift *art department lent me a professional 35mm camera, with which I demonstrated unequivocally and for all time that I was not, even under duress, a photographer.*

*Still, I interviewed everyone I could think of and took a couple of random stabs at scene reporting—loitering as the cruise ships were unloading, heading down to the yacht harbour at the south end of the island on the passing advice of one of my old journalism instructors (thanks, Lynn Cunningham!)—and came back with a genuine, full-fledged magazine piece. Well, most of one—*Shift *senior editor Joanna Pachner skillfully steered it the rest of the way.*

It actually grows more preposterous in retrospect that I started with this.

I. THE VIEW OF THE VIRTUAL VEGAS STRIP IS
OBSCURED BY FANNY PACKS

ON THURSDAY MORNINGS DURING high season in St. John's, Antigua, the big cruise ships come into port. There are tourist-laden boats at the downtown docks most other days, too—oversized yachts and multi-decked sailing vessels and even floating-hotel-sized ocean liners—but the *big* cruise ships come mostly on Thursdays. They are moored to the dock with ropes the diameter of an adult male thigh, and they transform the thin strip of concrete between them into a shadowed canyon like the business district of a very compact city. They tower eight or maybe ten storeys over the Antiguan capital's downtown core, easily dwarfing the tallest building in the city. There's an old unwritten law in Antigua that no building shall be taller than the tallest palm tree, and both ships currently in port are violating it. They are, far beyond the shadow of a doubt, what's *happening* in St. John's on this Thursday morning.

Around nine A.M., tourists begin to pour off the boats in an almost constant stream, thousands of them, turning Heritage Quay into a roaring river of walking shorts and tank tops, batik-print sundresses and gaudy T-shirts from yesterday's port of call, plus fanny packs, fanny packs and more fanny packs. Most of St. John's has come out to meet them—little pockets of cab drivers and handicraft hawkers and sellers of "island music" tapes, all of them planted like boulders against the rushing current. There are several steel-drum bands, and a Rastafarian banging on bongos, but the airwaves are dominated by booming loudspeakers set up in the back of a van, which broadcast a nonstop medley of cloying, calypsoed Western pop. The whole bazaar is thus set to a soundtrack of "Yesterday" and "Good Vibrations" and "Mr. Tambourine Man," all backed by "island riddims." At one point, in some strange, accidental synchronicity, "Auld Lang Syne" kicks in as a dozen costumed octogenarians wind their way through the crowd into a waiting taxi-bus.

Eventually, with the certain, irregular rhythm of a dissipating sun shower, the crowd disperses, some carried away in taxis and micro-buses, others swallowed up by the upscale duty-free shops along Heritage Quay. A few might even duck into King's Casino and lose a few hours and a couple hundred bucks at the stud-poker tables. But not one will notice The Sands of the Caribbean or Intertops Casino or Gold Club Casino or any of the other high-volume gaming parlours operating in St. John's. The tourists will return to the docks shortly before dusk, having lazed or shopped the day away, and their massive mothership will fade slowly away into the postcard Caribbean sunset, bound for St. Kitts or Puerto Rico or Martinique. And none of them will know that they've passed through the heart of the fastest-growing gambling town the world's ever seen, a controversial new Vegas built not on desert sand but on fibre-optic cable and liberal trade laws.

II. TEMPEST IN A LITTLE BLUE STREAMLINED BOX (TEAPOT SOLD SEPARATELY)

TO BE FAIR, IT'S NOT HARD TO MISS the casinos and betting parlours of St. John's. Take World Gaming, for example. To find it, you'd need to know that a drab, three-storey building on the edge of the down-town core houses the offices of World Gaming Services Inc., Softec Systems Caribbean and Electronic Financial Services Caribbean, all wholly owned subsidiaries of StarNet, a Vancouver-based soft-ware-development and online-content company. You'd need to head up to the second floor, wind your way past reception and through the cramped StarNet offices, then back through an empty room the size of a classroom to a small, windowless space that feels like a walk-in freezer (a comparison that would likely strike you immediately, since the place is air-conditioned to brisk-Canadian-November-day levels). There, sitting on folding tables, you'd find a mishmash of electronic gear, monitors and keyboards and circuit boards, plus three squat, cool-blue plastic boxes, each about the size of a watermelon. It is inside these boxes that, technically, all the action is taking place. Two of them are the servers for Worldgaming.net casino, plus StarNet's sports book and off-track betting parlour, plus twenty-four of the sites owned by licensees of StarNet's software. The third is the "bank," the place where the players at all these casinos set up accounts, buy chips and cash out when they've had enough. But even though you'd be standing right there, in the thick of the action, you'd still need a PC and net access to put ten bucks on black-twenty.

The more common way to make your wager—the way countless thousands of gamblers around the world have been placing their bets, turning online gambling into a $650-million (U.S.) industry in 1998 alone—is just to sit at home, fire up your modem, download the free interface and cash-transaction software (or have it mailed to you on CD-ROM for free), use your credit card to set up an account, and bet away at your leisure. That way, not only do you save yourself the cost of a Caribbean trip, you get around 250 different casinos and betting parlours to choose from, most with funky video-game-quality 3-D graphics that put the bare walls and techie clutter of StarNet's cold room to shame.

But regardless of which site you choose, the odds are still close to even that your cash will be heading to Antigua. Because even though there are sites operating from points around the globe, the majority have located their servers in the Caribbean. And even though the industry has established itself on other islands—including Dominica, St. Kitts, Grand Cayman, Guadeloupe and Curaçao—they are mere Renos to Antigua's Vegas. At last count, forty companies—nearly all founded by expats from Europe and North America—have set up shop in Antigua, incorporating there and receiving licences to operate casinos and sports books online. Most run at least two sites (say, one for casino games and one for sports betting, or else one casino with a Monte Carlo theme and one with a Wild West saloon vibe), but some have set up a half-dozen or more. All of which means that tiny

Antigua—twelve miles across and nine miles from top to bottom, with a total population (counting Barbuda, its sparsely populated sister island) of 64,000—controls somewhere between twenty-five and fifty percent of the booming online-gambling business.

And boom it has. It's too early to say that online wagering has changed the face of gambling—the industry will be doing well to break the billion-dollar mark this year, while offline wagering grossed an estimated $51 billion in 1997 in the U.S. alone—but it's definitely the biggest e-commerce success story this side of internet porn. Most online-gambling start-ups turn a profit their first year, and licensees of software from established firms such as Toronto's CryptoLogic (one of the industry's biggest players) are grossing, on average, $1.6 million every quarter. Then tack on the fact that the existing markets are expanding and virgin territory is being explored every day (many in the business are particularly giddy over the wiring of China, with its vast population and a reputation in the industry for fanatical gambling) and you've got a gold rush on your hands.

Antigua, like countless boomtowns before it, has grand plans, hoping online gambling can be used as leverage for a bold vault into the front ranks of the global information economy. Already, however, the island's online prospectors have caught the attention of the suspicious and litigious forces of American justice. Meanwhile, competition from other Caribbean islands and from elsewhere around the world is mounting. Some Antiguan-based gaming operators are even trying to set up satellite offices in Australia—all the better to hawk Pai-Gow poker and Mahjong to the lucrative East Asian marker. And then there's the small matter of getting a decidedly unwired island—a place where the home telephone line is still fairly newfangled—up to digital speed, and keeping it there.

In 1898, Dawson City, Yukon, was the "Paris of the North," the largest North American city west of Winnipeg and north of San Francisco, with phone lines, electricity and cancan girls; four years later, its population had dropped by eighty-five percent, making it just another forgotten village on the fringe of the industrial world. It remains to be seen whether the information revolution will be kinder to its Klondikes.

III. PHI DELTA EXILE (ANTIGUA CHAPTER)

SUNDAY IN ST. JOHN'S, EARLY AFTERNOON. The streets are slick with rain, a rarity on this arid island. Gold rush notwithstanding, the city has open sewers, and the air is tinged with the sickly sweet smell of offal and decay as I make my way to a downtown building called Ryan's Place, home of World Sports Exchange (WSE), an online sports book catering mainly to Americans. Today is game day—and I'm on my way to watch football with fugitives.

Haden Ware, a founding partner of WSE, is initially very matter-of-fact about the whole thing. "Oh, yeah," he tells me, "we've been indicted. We've actually had criminal complaints filed against us. We were very outspoken. I mean, we had zero customers when we first came down here, so in order to get customers we had to get our name out there. Jay [Cohen], our former president, was a very talkative guy. So we got in *The Wall Street Journal* and *The New York Times*, we were featured in *Sports Illustrated* and on some TV shows and stuff like that, and that basically got us our customers. But it was kind of a Catch-22. I mean, we got the customers, but at the same time we got a lot of press, so when the federal government decided to issue indictments and criminal complaints to some of the books downs here, we were on their list."

What were the charges?

Ware pauses, distracted. The computer on the table beside him pings once, twice, again—each ping an incoming bet. He's a big twenty-three-year-old, six-foot-plus and broad-shouldered, and he's restless and fidgety in conversations, like a college jock who'd rather be playing catch than answering questions.

"There were six counts of..." He pauses. When he starts again, the straightforward tone is gone, and he's over-enunciating through gritted teeth: "... of conspiracy to break the law and... I'm not sure if it was... *Steve?*" He calls to his partner, "Do you know exactly the charges that we've been charged with?"

Without looking up from his monitor, Steve Schillinger replies, "The Wire Act."

"The Wire Act," Ware echoes.

That's the Interstate Wire Act of 1961, more precisely, a dusty U.S. law that forbids the acceptance of bets over a telephone line.

Ware, along with Schillinger and Jay Cohen, who's left the company and the island to fight the charges in New York, was indicted along with eleven other online bookmakers in March 1998. WSE's lawyers are confident of acquittal, though none of the cases has yet gone to trial. "But as things are right now," Ware says, "Steve and I cannot go back to the States. It's kind of ridiculous. We're entrepreneurs. We came down here to start a business, and all of a sudden we're, we're fugitives. God knows why."

Ware's confusion is understandable, not least because far from feeling like a den of seedy bookies, WSE headquarters has the relaxed, slightly shabby coziness of a frat house. WSE inhabits a single room about the size of a high-school classroom, and is littered with over-stuffed couches and half-empty KFC containers. The walls are decorated with randomly placed sports pennants and posters. There's even a fridge stocked with beer and pop at the back of the room. Only the tables piled high with telecommunications equipment hint at the room's function, and their businesslike presence is toned down by the half-dozen college-age employees who slouch in chairs in front of the monitors. Even Schillinger, a grey and balding forty-five-year-old, comes across, in his T-shirt and flip-flops, more like some terminally youthful grad student than the boss of a small company.

But the kind of cash being generated by online betting simply isn't small. And while most governments have either adopted wait-and-see attitudes or are starting to look into reasonable regulatory measures (several Australian states, for example, have begun implementing licensing policies, and one Canadian MP has suggested that the Canadian government itself start a sports book and use the revenue to fund amateur sport), Washington has charged blindly toward prohibition. Back in March 1997, the Senate passed a bill, tabled by Arizona's Jon Kyl, to ban online gambling outright, but it stagnated in the House and has been labelled unenforceable by the Department of Justice. Meanwhile, crusading DAs in several states are hunting for career-making precedents under existing laws such as the Wire Act. And a bizarre lobby group that includes the Christian Coalition, several professional sports leagues, some of the owners of offline casinos, and Ralph Nader has formed to decry online gambling in general. The opposition is sometimes motivated by morality (*gambling is bad!*), sometimes by self-interest (*hey, that's our revenue!*), and sometimes by ethics (*they could use those binary whatzits to fix the games!*). Without fail, though, it ignores both the anarchic, borderless nature of the internet and the powerful human compulsion to gamble.

To wit: Back at WSE headquarters, it's five minutes to game time and the office's computers are pinging like cicadas in the Antiguan night as hundreds of last-minute bets, the vast majority of them from the U.S., roll in. Several employees are taking more bets by phone. At the centre of it all is Schillinger, who's double-checking the odds on the game that will be today's interactive-betting feature on WSE, and who could likely tell the feds a thing or two about gambling. Back in the States, Schillinger was deeply immersed in the culture of that state-sanctioned professional betting parlour called the stock market, working as a market-maker on the Pacific Stock Exchange in San Francisco for eighteen years before he relocated to Antigua. In fact, it was on the PSE floor that the sports-exchange concept was hatched. As a lark, Schilllinger started taking bets from fellow floor traders on just about everything he and his buddies talked about, from who'd win the Super Bowl to the verdict in the O.J. Simpson trial. And also as a kind of lark, Schillinger couched his small-time bookmaking in the grandiose terminology of his surroundings. That is, in place of odds, Schillinger installed futures prices; you didn't bet on your team, you bought shares in it. So let's say the 49ers looked unbeatable in the pre-season. Schillinger would set their asking price high—maybe 30 bucks a share—with a payout of $100 per share if they won the Super Bowl. Then let's say the Niners won their first five games—your stock would climb to, say, forty bucks, and you could sell right then at a tidy profit, just like some big-time broker.

Jay Cohen, a buddy of Schillinger's on the floor, thought the idea was custom-made for the net, so he raised about $1.5 million (U.S.) in venture capital to finance it as a legitimate business. In the fall of 1996, he, Schillinger and Ware (who had a summer job on the PSE floor as a runner) moved down to Antigua, whose government just happened to be the only one in the world offering internet sports-book licences.

WSE's website has never been much to look at: just a few frames, a white background and a bunch of multicoloured text. But

Schillinger's futures have found their niche online. The site offers sin-gle-game futures (for baseball, basketball, football, golf and, soon, horse racing) with share prices that change from play to play, and it's packed with off-the-cuff bets—who'll score the next touchdown and how, what a given ballplayer will do at his next at-bat—that give it the feel of watching the game with a case of beer, a wad of cash and your bet-crazy buddies all around you.

WSE currently has 5,000 such buddies registered and counting, and though Schillinger—like the heads of all except the biggest (read: publically traded) companies in the business—isn't willing to talk bottom line, he does tell me that WSE has been in the black since halfway through its second year. "You can make a good living at this," he adds with an impish smile that seems to confirm any fantasy I might by now be entertaining about relocating to the Caribbean and making easy money watching sports and handicapping the outcomes, hanging out with my buddies, making goofy side-bets and keeping four or five percent of the total take. As if on cue, two of Schillinger's employees start discussing a sailing trip they have planned for tomor-row. Another is messing around with some devil sticks. Just good clean fun.

Except that the wife and four kids standing with Schillinger in the framed photo next to his monitor are up in San Francisco, where he's a wanted man.

IV. ANTIGUA DREAMING, ON SUCH A WINTER'S DAY

"As LONG AS THERE ARE PEOPLE who want to wager, they will find means by which wagering can be done," says Gyneth McAllister. "The number of dollars spent illegally on sports wagering in the back alleys in the United States far exceeds what is being wagered legit-imately through the Caribbean. Here, they're just doing it in a safer environment."

That safer environment is Antigua's Free Trade and Processing Zone, a legislative invention of the Antiguan government. The zone was established in 1994 with much the same muddiness of purpose as most new-media start-ups: Antigua only knew it wanted to be involved in some kind of high-tech business or other. McAllister, the prime minister's Expediter for International Investment, quickly became a central figure in the zone's expansion (some would even say she was the online-gambling industry's architect), helping it to find its destiny as an infant industry grew up around it.

That Antigua's free-trade zone and online gambling found each other was a bit of a happy accident. It went something like this: Way back in the late '80s an organization called Sports International set up a telephone-based sports book in Antigua, taking advantage of the country's liberal tax and banking laws and the quasi-legal status of its operation on the island. SI chugged along until about 1995, when it moved online. Meanwhile, a flamboyant, self-aggrandizing Canadian (if you can imagine that) named Warren Eugene, a Grade 9 dropout from Toronto, was proclaiming himself "the Bugsy Siegel of the internet" to anyone who would listen, and started setting up net-based casino and betting operations on several Caribbean islands. Eugene got considerable press attention back in the States—*Wired* put him front-and-centre in a feature—precipitating what McAllister calls "a deluge" of interested parties making inquiries on assorted Caribbean islands.

"You see the predicament we were in?" McAllister asks me, smiling broadly. We're sharing bananas flambé, an Antiguan specialty, at an upscale bar and grill on Redcliffe Quay, a touristy shopping district one block from the cruise-ship docks. It's a glorious subtropical morning, all blinding sunlight, beach-ready warmth, air smelling of ocean, and McAllister can barely contain her glee. She is thirty-eight but could pass for at least a decade younger, with striking, high-cheekboned features, ceaseless enthusiasm and a deeply infectious laugh.

"So our legislation," she says, "went through some quick evolutions." Which is to say, the Antiguan government quickly threw together a regulatory apparatus inside the free-trade zone for phone-based sports books, then amended it to include casinos, then—the masterstroke—began issuing licences specifically for online casinos and sports books. The licences were a major enticement for nervous operators, especially from the U.S., who were, after all, getting involved in an industry traditionally associated with FBI stings and the ruthless politics of a Scorsese movie. They were particularly attracted to the fact that running an internet gaming operation was now explicitly legal on the island. (Many other islands merely allow the casinos to operate in "a grey area," as McAllister calls it, though some have followed Antigua's lead and are developing licensing schemes.) They were, of course, also lured by Antigua's decision to issue the licences under the rubric of the island's Free Trade and Processing Zone, meaning they'd pay no taxes on imported equipment and not a dime on corporate earnings, and their customers would pay no taxes on their winnings. All this for an annual licensing fee of $100,000 (U.S.) for casinos and $75,000 (U.S.) for sports books.

Which brings us to online gambling's allure for the predominantly unwired people of Antigua, and to the daring socioeconomic experiment on which Antigua has embarked: Can a nation that in many ways never completely caught up with the latter part of this century—a nation with a per-capita GDP of $6,800, a place with open sewers and roads that range from merely pockmarked to lunar-landscaped—catapult directly into the next? Or is this, as it might seem at first glance, just another wave in Antigua's historic colonial relationship with the developed world?

Antigua has never been anything but a natural-resource base for the big imperial powers, and a minor one at that. From the late 1600s up until about twenty years ago, it was one of the smaller of Britain's numerous Caribbean sugar-plantation colonies. Then, coincident with (and partly because of) the rise of a trade-union movement that eventually led to independence in 1981, the sugar industry collapsed. Antigua today is littered with squat stone sugar mills (they often show up on postcards and brochures) and less-photogenic aluminum-sided refineries, all abandoned. Following the lead of its Caribbean neighbours, the country then jumped wholesale into the tourism industry, building an international airport (partly paid for, incidentally, by the Canadian government), dozens of resort hotels and the posh Heritage Quay duty-free district. Its marketable natural resource changed from sugar to the fine white sand of its beaches. And now the casino operators have come, drawn this time by the island's libertarian attitude toward gambling, taxation and personal and corporate finance (banking laws are roughly equivalent to those in, say, the Caymans, and the government is hoping that net-based offshore banking will be the country's next big industry).

This time around, however, Antigua has no intention of being used. In August 1996, just as online gambling operators were pouring into the country, Gyneth McAllister went to Washington. Antigua couldn't have asked for a better negotiator: McAllister is Antiguan by birth but attended one of Georgetown's toniest girls' schools, and worked in Washington for a time in the office of her father, a lobbyist for the U.S. oil industry. McAllister met with high-ranking officials from the departments of State and Justice, where she presented Antigua's gambling legislation for the scrutiny—but not censure—of Antigua's overbearing northern neighbour. "I opened by saying, 'I am not here to be bombarded with reasons why the industry should not exist. The industry does exist, it's always going to exist, so let's move on from the tantrum stage. Let's do something that you're going to find really hard: Let's actually give us some respect here and treat us like equals and work side by side, rather than looking down and patting us on the head and giving us instructions.'

"Now that," McAllister says, "is a huge jump, a *huge* evolution." And notwithstanding the indictment of a few American citizens, the American government, she says, has since been cooperative rather than dictatorial, no matter how hysterical certain senators and DAs have become.

But even more important to Antiguans than the prevention of U.S. meddling have been a couple of provisions in the aforementioned legislation that ensure that the island's people reap some benefits from the industry's success. Foremost among these is the Free Trade Zone Institute of Technology & Training (FTZIT&T), an IT school financed by the licensing fees paid by gambling operators. The school, a converted furniture warehouse on the outskirts of St. John's, isn't much to look at, but the classrooms inside are stocked with late-model Pentium computers. And after only a year of operation, about a thousand Antiguans (mainly government employees) have received training in HTML, Visual Basic, C++ and computer assembly and repair. The school's director, Patrick Lay, says the government intended to train its employees even before the gambling bonanza. But once the licensing fees started rolling in, the initial plan to farm the training out to a California firm was scrapped in favour of building the FTZ-IT&T. As well, under the licensing agreement, net-gaming operators are required to employ four locals for every foreign national working in their Antiguan office, meaning that the industry is providing further training to FTZIT&T grads or, as is more often the case, training its local employees from scratch. And while a single school and a few hundred jobs at high-tech firms might seem trivial elsewhere, in a nation of 64,000, it's enough to have everyone involved talking dreamily about the future.

V. CAREER OPPORTUNITIES

"This is a window in your life," Gyneth McAllister is telling me, leaning across the table, pushing aside the remnants of our bananas flambés. She has spent the morning explaining to me how online gaming works. She has outlined Antigua's regulatory apparatus—which includes background checks on the operators and their boards of directors, as well as auditing of software by an approved outside source (usually PricewaterhouseCoopers, an internationally respected accounting firm with a large office in Antigua)—making a strong argument that it is second to none. She has grown giddy describing the opportunities presented by the emerging Asian markets, saying, "Do you know what this means? This means being emperor of the universe, essentially. Now I find that appealing." In fact, she gushes, none other than Stanley Ho— the billionaire financier who built the Portuguese colonial outpost of Macau into the Hong Kong of Pacific-Rim gambling—visited Antigua recently, scouring prospective sites for his next venture after Macau returns to Chinese control later this year. She has told me all this, and now—she can't help herself—she's pitching me on a licence.

"This is a window in your life," she says. "It's not often that you're walking down the street and opportunity hits you in the head like *The Beverly Hillbillies*." She pauses, her grin widening to a point where it seems to defy certain basic tenets of human physiology. "You've just fired your shotgun. The bullet has pierced the ground, and the black oil is shooting up like a fountain. What are you going to do?" Another pause. "You see how I get people into the industry?"

I answer yes, and start pondering whether anyone I know has access to venture capital.

So let's say I decide to make a go of it. The first thing I'm going to need after a licence is software—both the online interface and something to make those easy millions in cash transactions secure at both ends. And unless I'm already involved in a high-volume e-commerce industry—like Vancouver's StarNet, which built its transaction software in the adult-site business—or I have the time and money to put together a team of programmers, designers and especially cryptographers to build my software from scratch—like Toronto's CryptoLogic did—then I'm going to have to license someone else's software.

Now, it'll cost me: Complete packages from the established players (StarNet, CryptoLogic, BossMedia of Sweden, MicroGaming Systems of South Africa), which include a customized interface, cash-transaction software and optional add-ons like maintenance of my site and my customers' accounts, run to about US$100,000, plus around thirty percent of my revenue. But in the long run, the investment will likely be worth it, because the software—especially the cash-transaction part of it—will be the cornerstone of my business; no one, after all, is going to bet with someone who can't guarantee that their account is safe. And the gambling industry (along with its close cousin, pornography) is at the cutting-edge of e-commerce technology.

VI. WAITING ON THE TRICKLE DOWN

"YOU SEE THAT?" MY CAB DRIVER SAYS, pointing to a building about the size of a hockey rink. "Sugar factory. Abandoned." The factory slumps forlornly on the roadside like an exhausted hitchhiker, looking as if it might collapse completely at any moment. We are heading south out of St. John's, with its tiny, tidy bungalows and corrugated-siding shacks, through rolling fields populated by scrub trees and the occasional village. We are bound for the yachting enclave of English Harbour on the island's southern tip, and the cabby is graciously pointing out the sights.

"There," he says a little while later, indicating a squat, cylindrical stone structure that looks like a decapitated windmill. "Sugar mill. Old style." Also abandoned. As a young man, my cabby worked in the cane fields; everyone from his village did. Then the industry fell apart, victim of a Caribbean-wide rationalization effort in the seventies that had no place for the few ancient plantations on tiny, arid Antigua. The island's rum, Cavalier, is now made with imported sugar cane. And men like him, who used to work the fields, now drive taxis or clean hotel rooms or ring in duty-free sales.

It's a short drive to English Harbour, maybe half an hour along narrow roads, slow-moving and full of dramatic swerves as yet another small canyon of a pothole is narrowly avoided. But it's a world away from the bustle and squalor of St. John's. On this day, there is a boat show—THE NICHOLSON ANNUAL CHARTER YACHT SHOW, a banner proclaims—and the harbour is packed tight with sleek forty-foot yachts and sailing vessels that look ready for the Americas Cup. At the picnic tables outside the dock canteen, tanned jet-setters with assorted European accents sip beer and renew old acquaintances with the unself-conscious, casual confidence born of generations of aristocratic privilege. The dock is *theirs*, and they know it. I wander off and find a seat on a rock along the harbour's shore, thinking of Robert Vesco.

Vesco brought *his* sleek yacht to this harbour in late 1981 after ten years in hiding in the Bahamas. He had fled the U.S. in the early seventies with $224 million (U.S.) worth of investments in his fraudulent mutual-fund company. Antigua was newly independent, and in its first—and, to date, most high-profile—defiance of the American government, then-deputy-prime minister Lester Bird (the son of the then-prime minister, V.C. Bird, and now himself PM) allowed Vesco to stay for six months at English Harbour despite American pleas to help bring him to justice. This and subsequent affairs have earned the Antiguan government a reputation for corruption, echoes of which could be heard in news reports about the failure of an Antigua-based internet bank and the disappearance of its two founders in the fall of 1997.

Add to this the gambling industry's historic ties to organized crime, and the American opposition to its move online begins to make sense. Even with the Disneyfication of Las Vegas in recent years, gambling is still a suspect business, inherently not on the level. Which explains why the American government officials who met with McAllister were the ones in charge of money-laundering and computer fraud; and why

CryptoLogic's stock hasn't taken off like its founders had hoped; and why Haden Ware and Steve Schillinger were pursued so doggedly by a New York DA. Because these are bookies and pit bosses we're dealing with, and those sorts of people just have to be criminals, right?

But the net's gambling entrepreneurs just aren't that kind of mob: Terry Bowering, vice-president of StarNet's offshore operations, is a former stockbroker from Saskatchewan; one of my meetings with Gyneth McAllister was mediated by the sound of her pre-teen kids watching cartoons on my hotel TV; and World Sports Exchange's "fugitives" all but ooze all-American wholesomeness.

Maybe, I thought, as I watched the shadows of the masts grow long on the harbour, maybe this wholesomeness is why the Antiguans involved in net gambling are so enthusiastic, so full of evangelical zeal about the promise of the internet, the computer age, the future. Here was an industry tailor-made for their nation's *laissez-faire* approach to the global economy's rough edges, but at the same time a business that didn't intend to rape the island of its few natural resources or sequester itself in opulent enclaves. An enterprise that merely wanted to be on the island; that merely wanted, moreover, to *be*. For Antiguans, here, finally, was an opportunity not only to meet the world's economic giants on equal footing, but even to lend an air of legitimacy and professionalism to one of their most unsavoury pastimes. Antigua, as many of the net's casino operators told me, has a reputation in the business—a reputation for keeping it fair, honest and above board. It's a reputation the country undoubtedly wants to encourage.

As I sat on my rock, lost in thoughts along these lines, a Rastafarian took a seat next to me and introduced himself as Byron. After some small talk, he asked me if I'd buy him a beer. I went to the canteen and bought us each a Red Stripe, and we sat and talked as we waited for the sun to set over the flags and on the harbour. Byron wanted to know where I was from, why I was in Antigua. I told him I was a reporter from Canada, but didn't elaborate; online gambling and the promise of a wired future for the nation suddenly didn't seem relevant. I asked him where he was from. He pointed to the hills to our left. "My whole life I have lived there," he said.

A whole life lived on the edge of the yachting circuit, watching the world's elite treat his backyard like its own private country club.

I wondered whether the countless millions being bet in the world's wealthiest nations and processed by servers on this island would ever trickle down to Byron—whether net gambling would be any kinder than the big-boat network. The sun set and our beer finished, I got up to leave. Byron asked me for a light for the joint he had rolled. I handed him my lighter and told him to keep it.

"Is it from the heart?" he asked, holding it to his bare chest. "I'll only keep it if it's from the heart."

Well, *damn*, Byron, that's a hell of a question. *Was* it from the heart?

I told him yes, and he kept the lighter, but I should have told him I'd get back to him on that when I could send the answer to his email address. Because—let's be honest—until then, I can't be sure.

VII. MEANWHILE, BACK AT THE FRAT HOUSE

STEVE SCHILLINGER HAS SILVER-BLUE EYES that dance with the lively energy of a kid on a playground daring you to take on the school bully. Tonight, those eyes are fixed on a Thursday night NFL game on one of the two TVs on the table in front of him. His hand is poised on his football-shaped mouse, ready to change share prices on the computer to his right. The Philadelphia Eagles, who have been looking strong since the opening kickoff, complete a big pass, a thirty-yard gain. Schillinger considers it for a moment, then bumps their stock up five bucks. Then he waits—a couple of pings from the computer, no more.

Later, with Philadelphia in complete control and the third quarter winding down, Eagles stock breaks the ninety-dollar mark. *Ping. Ping ping ping-ping. Ping ping.* A number of the bettors who've been following the game, and the market, all night finally dump their worthless St. Louis Rams shares, taking the loss on the chin.

The Rams have the ball now. A first down, then another. Philly's lead is only ten points, with almost a full quarter to play. *Ping ping ping. Ping ping.* A few bettors—including some who already have Eagles shares—scoop up the dirt-cheap Rams futures. Hey, you never know.

Then: touchdown, St. Louis! Schillinger's computer seems about to leap off the table with the force of the pings. He bumps up the price of the Rams until the bets subside. This is how Steve Schillinger, once of the PSE, makes markets these days.

It's a quiet night at WSE—just Schillinger and a techie who's making sure the bets are being recorded and paid out at the right time, and that the odds for the impulse bets are changed after every touchdown. The Philadelphia–St. Louis game is the sole interactive tonight, and the only other action is on a couple of college basketball games. Still, the computer rings out pings, and the telnet window I've been using to watch the bets keeps filling up, then scrolling further down to make room.

After every touchdown, another round of impulse bets pops up on the screen: six, ten, sometimes more. I've been watching someone with the handle "havno," who Schillinger told me was a doctor in Arkansas or a lawyer in Mississippi or something (many of his customers call in the bets when their net connection is too slow or whatever and he's gotten to know some of the bigger ones). Philadelphia field goal—havno bets a hundred bucks on St. Louis to score a passing touchdown and $20 more on a kickoff return. St. Louis touchdown—havoc drops $300 on another Philadelphia field goal, plus ten bucks on a Rams interception return, a long shot with a big payout.

In the end, Philly wins 17–14, and WSE is up about $4,000 on the interactives. This is good enough for Schillinger, who usually turns a small profit but has "messed up" to the tone of $30,000 or more on a couple of games. And, after all, it was a nothing game on an off night, a late-season match between two teams with not even the ghost of a chance at a play-off birth. And yet all night *ping, ping ping, ping ping-ping ping.*

I think then of a story another gaming operator told me about a Malaysian customer, a guy who only bets in the five digits. The Malaysian was sending a bet of $100,000 on a World Cup soccer game, and called up to see if he could maybe get a $10,000 bonus on it. The operator told the Malaysian he couldn't do it, but offered him tickets to the game instead. "I don't want to watch the game," the Malaysian told him. "I want to bet on it."

I think of how this operator laughed at that, and how he told me, "It's still a virgin market, it's still untapped, and there's more than enough for everybody." And WSE's Haden Ware calling online gambling "a growing industry" and StarNet's Terry Bowering saying it's "in its infancy" and Gyneth McAllister talking about how the main problem operators are having is that many of them "have never had this amount of money to deal with before."

There are estimates that online gambling will be a $7-billion (U.S.) industry by 2001, or maybe $10 billion by 2002. Or both. And I think of havno sitting in his comfortable living room somewhere in the American South, dropping $200 and $300 and more on who might score next in a meaningless Thursday-night football game, and I wonder if these estimates aren't just a little conservative.

TAKE ME DOWN TO PARADISE CITY

Shift, June 2000

The best happy accident of my career, and still moment for moment and word for word maybe my personal favourite story. The most fun I ever had on assignment, in any case.

I was on a stopover in Thailand, en route from India to Vietnam to spend the Christmas holidays with friends in Hanoi. I walked into a Bangkok internet café, logged onto my email account, and found a message from a Shift editor (Maryam Sanati) asking if I was anywhere near Malaysia. The Malaysian prime minister, she explained, had gone off on some crazy infrastructure project called Cyberjaya. Was I up for some field reporting?

I took an overnight train from Bangkok to Kuala Lumpur. I got lost in the palm jungle south of the city more than once. I ate my weight in mangosteens (which I wrote about for Maisonneuve *a couple years later). I transcribed several interviews with the powder sand of Langkawi warm between my toes. In my weaker moments as father and husband and mortgage bearer, I still wish just a little bit that the whole rest of my career had been like this, though I'm old enough now to know perfect impossible ridiculous Southeast Asian assignments can only ever form the delectable exception to the rule.*

Have a look at my mousepad here. WHERE HIGH TECH AND PARA-
DISE MEET, it says. Check out the slogan on my room key. THE MODEL
INTELLIGENT CITY IN THE MAKING, it reads. CYBERJAYA, it adds. Which
is where I am.

But then ask at the front desk why my in-room computer won't log
on. "System," the clerk explains, "is not working." Which is to say, "The
server is down." *Model intelligent city?* Maybe. *In the making?* Definitely.

Here's the thing: I'm staying at Cyberview Lodge Resort, that
newly fabled intersection of high tech and paradise mentioned on the
mousepad. To get here, you'll need to fly to Kuala Lumpur, Malaysia
(known around here as KL). You'll touch down at KL International
Airport, rumoured, at least locally, to be the most modern in the world.
Could be: It's all sleek lines, glass and matte-grey metal, sunlit atriums
and automated "aerotrains" to whisk you from terminal to terminal.
Find a cab and tell the driver, "Take me to Cyberjaya." (Malaysia was
a profitable raw materials outpost of the mighty British empire until
1957—rubber plantations and tin mines, mostly—so there's no need
to consult a phrase book; he'll speak English.) Off you'll go, down spacious and seamless new expressways, into what feels like very orderly
tropical wilderness: row after ruler-straight row of palm trees—dense,
but orderly. Palm-oil plantations. You'll move quickly down those new
roads—past the odd bungalow on low stilts, a jackfruit stand or two, a
gas station—until you're pretty sure you've come to Nowhere.

Just about then, you'll start to see clusters of low buildings through the palms. Little hypermodern pods of office space. Cyberjaya. Pull off the main road, up a low, landscaped hillside, into one of the larger pods, a place that feels like the offspring of a suburban office park and a Hawaiian golf resort. Welcome to Cyberview Lodge. Pick up your smart card at the front desk, find your room and pop your card in the slot labelled E-SWITCH just inside the door. The lights will come up—not glaring, just right—and the A/C will start blowing, and the computer on the desk will come to life and run a short A/V orientation thing for you. Never mind the V, but check out the A:

> Create a vibrant vision of the future city. *Place it*—in a verdant location of natural beauty. *Harness it*—with leading-edge IT and multimedia networks, infrastructure and world-class companies. *Protect it*—with eco-friendly resources and strategies. *Cherish it*—with nature sanctuaries, cultural parks and tropical gardens. And we have. *Cyberjaya*: Malaysia's first model cybercity in the Multimedia Super Corridor—a futuristic intelligent city, born of the people's vision and collective will... To extend a global invitation to celebrate a creative, energizing [*unintelligible*] in a global community of creative enterprise, industry, recreation and choice lifestyles of the future...

Got it? What the warm, vaguely Aussie-accented, vaguely Orwellian woman's voice is telling you is that you've arrived in the future-is-now centrepiece of Malaysia's Multimedia Super Corridor. Now, the MSC is a simply colossal public-works project (estimates of the price tag vary from $2 billion to $40 billion), stretching north in a nine-mile-wide band for thirty miles from the über-modern airport to the highest-in-the-world twin peaks of the Petronas Towers (88 storeys, 1,483 feet) in the heart of Kuala Lumpur's twinkling city centre. It includes Cyberjaya's sibling city, Putrajaya—the new administrative capital of this ambitious nation you're in. And it also encompasses some lower-profile initiatives. There's a state-of-the-art telecom infrastructure—which includes a fibre optic backbone with a 2.5 to 10 gigabyte-per-second capacity—and a whole mess of perks designed to attract information-age companies (everything from five- to ten-year tax holidays and duty-free equipment imports to removing restrictions on foreign ownership and the employment of "foreign knowledge workers"). There's a handful of recently enacted "cyberlaws" (copyright protection for intellectual property, legal recognition of "digital signatures," that kind of thing). And there's a government-owned corporation—the Multimedia Development Corporation—whose sole aim is to make you and your IT business happy. What's more, there's Cyberjaya's brand-spanking-new Multimedia University, whose primary (if not sole) aim is to provide you with well-trained knowledge workers. And then there's … well, you get the picture.

This, then, is the rough geography of the MSC, the brainchild of one Datuk Seri Dr. Mahathir Mohamad (Dr. M for short), Malaysia's prime minister since 1981. The MSC is the wannabe Silicon Valley of the East, the would-be pole with which Malaysia and its grandiloquent PM intend to vault as fast as possible from ailing Asian tiger to developed nation.

Got it? Good.

But here's the thing: The server is down. Cyberview Lodge is offline.

So. Here I am, "at the heart of the Multimedia Super Corridor," to quote my mousepad once more. A stone's throw from MSC headquarters. A healthy Frisbee toss from Cyberview Garden. A light five-iron from Cyber Heights Villas. A leisurely mountain-bike's pedal from Multimedia University. At pretty much ground zero of what is—if the Malaysian government has its way—the Next Big Thing in IT. I have emails to send—say, to the Malaysian reps of some of thirty-three blue-chip companies the MSC has attracted to date (a list that includes Lucent, Microsoft, Motorola, Nokia and Sun Microsystems). But the server is down. I have contact numbers to track down—everyone from Malaysia's energy, communications and multimedia ministers (yes, Malaysia has a multimedia *minister*) to the head of Mesdaq, the nation's one year-old baby Nasdaq. And I could be doing more research, of course. Why, for example, does Alvin "Future Shock" Toffler—one of the lesser lights on the MSC's International Advisory Panel (a who's who that includes Bill Gates, Larry Ellison, Scott McNealy and Lewis Platt)—why does Toffler think that Dr. M's vision "seems to have evaporated"? Who knows? Not me: The server is down.

"Maybe tomorrow," I'm told at the front desk (the very same front desk, incidentally, that was a little bewildered when I explained that I'd made my reservation at their hotel online). "You try tomorrow morning."

I HAVE SOME TIME ON MY HANDS. So I buy drugs. Of course, this is Malaysia, where huge billboards inform you that drug trafficking is punishable by death. So I have to improvise: I head over to the Cybermart across the parking lot and buy an armload of Tuborg beer, an equal measure of Red Bull (an "energy drink" with copious quantities of caffeine, plus large doses of "taurine" and "glucoronolactone") and a pack of Indonesian clove cigarettes. And then, in the shadow of my offline computer, I settle in. I listen to the drone of the A/C and tropical insects, which coalesce, after a couple of Red Bulls, into a sound kind of like ambient techno or, well, something up-to-the-minute, anyway. I read articles, flip through notes and take a swim in the pool. I spend some time in the lobby, staring spellbound at the photorealistic video aquarium that blows simulated bubbles by the door. (Those sure look like *real fish*, I think to myself.) I wonder, for a time, how far I'd have to walk north through the palm plantations before I found an all-night cybercafé.

Finally, near dawn, after I've counted all the layers of irony contained in being offline in Cyberjaya, I retire to my balcony and stare into dense tropical fog. And I wonder—can you blame me, at this point?—if maybe this whole MSC thing isn't a bit far-fetched. And right around then, like the song says, somebody spoke and I went into a dream . . .

THE VOICE IS SOFT, BUT SURE. "Sometimes," it's saying, "it is necessary to jolt people out of their complacency and smug self-satisfaction." And now, whaddaya know, out of the fog comes Dr. M himself, seventy-four, weathered of face and a little jowly, but with hair still jet black and eyes still lively. There he is—the Ghost of Malaysia Future—and the voice is his. Turns out this is more a vision than a dream. Dr. M's vision: Vision 2020. He settles into the chair next to me on the balcony and continues: "We need to change our perception altogether toward our environment, society, the way we rule and the way we do business. We cannot go back to invent the wheel. We have to leapfrog into more modern science and technology."

This is the same kind of thing Dr. M was telling the Malaysian people back in the mid-eighties, as he set about pushing the nation's emerging industrial economy (which began in the early seventies with a few foreign electronics companies) into overdrive. In 1983, he launched the national car company, Proton. By the nineties, Malaysia had become the world's largest exporter of semiconductors, riding a wave of unprecedented growth that made it the globe's second-fastest growing economy (after China), and the developing-nation darling of hedge-fund managers the world over. It was right about the same time that Dr. M started talking about Vision 2020, a slogan that you'll now find proudly displayed on the rear windshields of a great many Proton cars.

"To become a developed country, according to our Vision 2020," Dr. M explains, "we need to tap the talents of the whole world." Convinced that Malaysia had to act—and now—to get in on the information boom, Dr. M took a two-month working holiday, and spent most of it talking to the elite of Silicon Valley. He also commissioned a major study (from management-consultancy behemoth McKinsey & Company) to determine what kind of environment the information industry required. The result, announced in 1996 and launched the next year, was the MSC.

Dr. M is sweeping his arm gently toward the fog now, beckoning me to follow. And then we are off through the palms...but there are almost no palms left, just a few neat rows of them, decorously lining the highway that runs through the heart of Cyberjaya, linking it swiftly and directly with both the airport and downtown KL. And here we are, with the highway in the distance, walking a busy promenade—some kind of carefully sculpted garden boulevard, Cyber Axis by name, a newer, greener Champs-Élysées bordering a man-made lake—that ends at a central square. (And are those *families* we're passing—happy, healthy knowledge-worker families—out for an evening stroll?) Over there is a gleaming office tower, and suddenly we're *in* it, on the observation deck of Cyberjaya Tower, and all around us is a bustling city of cozy neighbourhoods and smart schools and zero-emissions vehicles, office parks, wetland sanctuaries, mosques and even a golf course by the university. Far in the distance—Dr. M pointing proudly, directing my gaze—is some kind of digital Disneyland, a big theme park, holiday resort and also working multimedia studio (set up under the tutelage of the United Kingdom's Leavesden Studios, where *The Phantom Menace* was made way back when). And even farther off, Cyberjaya's prosperous sprawl bleeds into Putrajaya's, both merging seamlessly with the older suburbs, and the whole thing spills into downtown KL, with the Petronas Twin Towers a final double exclamation point on the horizon.

"I see the MSC having hundreds of large and small companies working collaboratively with one another and with partners across Asia," Dr. M says now. "The MSC will be a global community living at the leading edge of the information society." The smart city laid out before me—population circa one hundred thousand—is linked to dozens of others elsewhere in the KL area, throughout Malaysia and around the world.

And then we are back at Cyberview Lodge, and Dr. M stands again at the edge of the fog. "What I have just described," he says, "has probably never been attempted anywhere else in the world. You may be thinking, 'Why Malaysia?'" And then he's gone, leaving me to wonder. And to wait, until late morning, for the server to come online.

ROAD TO NOWHERE

HERE'S SOMETHING ABOUT THE HEART of the MSC: It is, at present, a touch remote. Not just from KL (which is at least a half hour away by car), but even from itself, from its constituent parts. Vast walls of palms separate Putrajaya from Cyberjaya, and Cyberjaya from the airport, and even the scattered pockets of nascent Cyberjaya from each other. This is what I'm thinking, as I pedal my rented mountain bike down the road from Cyberview Lodge in the general direction of some construction cranes on the horizon.

It seemed like a clever plan: Ride a mountain bike around Cyberjaya, just like all those hip, young knowledge workers the city hopes to attract. Get a feel for the place. But I hadn't taken into account the thick, tropical heat, and I'd definitely misjudged its size. I push on, past newly paved side roads that run a hundred yards off the main road and then end abruptly in brick-red dust and swamp; some day, these will wind into neighbourhoods and commercial districts, or become avenues lined with trendy cafés and boutiques. Some day. (David Byrne's voice in my head now: "There's a city in my mind / Come along and take that ride / And it's all right, baby, it's all right / And it's very far away / But it's growing day by day / And it's all right, baby, it's all right," as I cross the intersection of yet another hundred-yard road to Nowhere.)

There appears a vision on the horizon: a low knoll topped with clusters of buildings, roofed in ocean blue. It is the campus of Multimedia University, built from scratch in eighteen months by construction crews working around the clock, and now home to thirty-four hundred students (with an eventual target of twenty-one thousand). I pull off the main road, and as I crest the hill, the call to prayer begins to ululate over the campus from the university mosque. It gives the place a timeless, dreamlike feeling, almost as if it were haunted—the Ghost Town of Malaysia Future. The feeling deepens as I circumnavigate half the campus without seeing a single person. From the hill, I can see the skeleton of Cyberjaya stretching outward: bands of dense palm forest alternating with wide ribbons of freshly turned red earth (the latter populated by construction equipment), buildings half-encased in scaffolding, the outlines of construction cranes towering over the trees in the distance.

I come across a sign—CYBERCAFÉ TAU—and head that way for a drink, past student dorms and the MSC Central Incubator (not as *Brave New World* as it sounds: an office complex offering dirt-cheap rent and expert advice for fledgling multimedia start-ups). I traverse a parking lot, hop my bike over a curb and cross the main quad, in the direction of a Pepsi machine. I've nearly reached it when a voice calls out: "Hey! *Hey!* Ehscuse me ... *suh* ..."

I turn to find a middle-aged Malay man striding across the quad toward me. He's pointing vehemently back toward the parking lot, but seems more surprised than angry.

"No bike," he says when he reaches me. Confused, I climb off the bike, smile and nod, and then begin wheeling it toward the Pepsi machine.

"No ... *no* ..." He taps my arm. "No bike here. You must leave over there." He points to the parking lot again, and this time I notice a line of motorbikes at the edge of the parking lot.

"But this is just a bicycle," I say. "Not a motorbike."

"No. You leave over there." He smiles—this is not a demand, but a friendly suggestion—then adds, "You could get fine." And then he stands there, waiting for me to comply. I turn and wheel the bike back toward the parking lot, and it occurs to me as I do so that I'm probably the very first person ever to bring a mountain bike to the MMU campus.

As I obediently—and somewhat irritatedly—walk across the quad, I notice a sign on a building to my right: FACULTY OF CREATIVE MULTIMEDIA MANAGEMENT. Now, how, I wonder, do they expect to teach "creative multimedia" at a school where you can't even ride a bike across the campus? I find myself growing skeptical of this whole MSC endeavor. But it is a tangential skepticism, a sort of dubiousness by association. And it is, I've noticed, the primary mode by which the MSC's critics voice their misgivings. Apart from the sheer improbability of the MSC—beneath suggestions that a nation that's just beginning to recover from a brutal recession cannot afford it—there is an unspoken belief that Malaysia is just not *ready* to become an information hub.

Consider Alvin Toffler's aforementioned about-face on Dr. M's clarity of vision. His attack came in November 1998 (briefly: "Instead of a sound strategy to compete in the global information age, he seems to be adopting the tactics of a police state"), not long after Dr. M sacked his deputy prime minister, Anwar Ibrahim. Ibrahim was subsequently arrested on corruption charges, then appeared in public for the first time after his arrest with a black eye and looking generally roughed up, and was later sentenced to a six-year prison term. The arrest, in Toffler's mind, proved that, for all his bright ideas, Dr. M did not understand how a responsible citizen of the global village was supposed to behave. His name is on the Committee to Protect Journalists' list of Top Ten Enemies of the Press, for chrissake. Now, how could a man like that expect to be welcomed into the progressive circles of the digital world?

Note, though, that Toffler wasn't quibbling with the fundamentals of the MSC plan. His kind of skepticism is aimed much more broadly at Malaysian society itself. Societies with authoritarian leanings can churn out tin and electronic widgets to their hearts' content, goes the reasoning, but they aren't allowed to play with the really cool toys until they learn some manners. After all, doesn't the phenomenal success of Silicon Valley (the model for the MSC) owe no small debt to its limitless freedoms—to toss around venture capital like confetti, to speculate on dodgy tech stocks, to let knowledge workers dress in dirty T-shirts and pierce their lips and even smoke a little pot if they need to before sitting down to design a website?

Now, how does this Mahathir fellow expect to attract hip, young knowledge workers bursting with innovative ideas to a place with limited press freedoms, draconian drug laws and all those autocratic tendencies?

There is a bigger factor lurking in the shadows of this argument. It's an uncompromising beast, and its name is Globalization. It wears fancy Western duds—equality, prosperity, all the finery of the free market—and it tells its developing-world "partners" that it's come bearing a whole new economic order. There are new rules, it says—rules that give everyone a fair chance; it might even talk about "level playing fields" and "win-win situations." But this same Globalization—which claims to detest top-down, hierarchical systems like Dr. M's—is something of a top-down, hierarchical system itself, and it would like Malaysia to abandon some of its bad habits before it'll be allowed to move up another rung.

Dr. M, though, has never been one to buy into Globalization's common wisdom wholesale. "The developed countries extend their assistance but, at the same time, [take] advantage to dictate terms to the poor nations," he's said, regarding the First World's role in Vision 2020.

Back on my bike, finishing my circuit of the MMU campus and thinking maybe Dr. M doesn't "get it" after all, I fail to realize that *I'm* trying to dictate terms. I'm just thinking that I'm pretty sure *I* wouldn't want to go to a school with a mountain-bike ban, and maybe that means I wouldn't much like this model city Malaysia is building. Presently, the road slopes up to the crest of the hill on which the campus rests. To my left, the land tumbles away down a grassy slope, ending in dense palm forest. To my right is a cluster of six- and eight-storey buildings, white-washed and tidy: dorms. At the base of one of them is an open-air cafeteria, and a dozen or so students—*people!*—are hanging out at the tables. I stop and watch these Malaysian college kids, imagining an idealistic, irreverent gang of future multimedia artists and designers. And then, amid a mishmash of stereotypical college garb (jeans and a rainbow assortment of T-shirts) hung out to dry, I notice a single poster taped to one of the windows of the building nearest me. A campaign poster. *Aha!* I think, pedalling up for a closer look. There is a ban on political demonstrations on campuses in Malaysia, and some young firebrand has bravely violated it. But the face on the poster is Dr. M's.

Fight the power.

I bike back to the hotel in sweaty, befuddled silence.

SUNDAY, BLOODY SUNDAY

A FEW NIGHTS LATER, JUST AS Saturday night is giving way to Sunday morning, I find myself following a BBC film crew through KL's Planet Hollywood at a party for the MTV Europe Video Music Awards. The crew is interviewing partygoers at random on Malaysian politics, because Monday is election day, and because (I suspect) the foreign press, in general, has been portraying this national election as a referendum on the future of Malaysia, pitting Dr. M against a youthful, reform-minded opposition that has united behind Anwar Ibrahim's wife in the wake of his arrest. Planet Hollywood feels like a campus pub—lots of hip Malaysian youth, lots of random exuberance and sexual innuendo on display—except the crowd is far more polite and way better dressed. Anyway, the point is that just as Sunday—election eve—dawns, U2's Bono is up on the big screen, receiving some sort of award for his years of selfless humanitarian activity. And, of course, he starts talking about Jubilee 2000—an effort, for which he has become the de facto spokesperson, to get the world's wealthiest nations to forgive the debt owed them by the world's poorest.

At this moment, the crowd near the bar erupts in boisterous cheers... as all eyes are fixed on a young Chinese-Malaysian woman, her head tilted back onto the bar, knocking back a shot poured from a bottle stuffed down the front of a bartender's pants. And it occurs to me, once again, that Malaysia's youth might not care too terribly much that certain foreign observers consider its government too autocratic for the global village.

But I don't want to make assumptions. So I accept an invitation from Shariran Shaari, or simply Sha, who is thirty-seven and runs a local cybercafé, to see what he describes as the region's most contentious electoral district. We arrive in Bangsar, a neighbourhood on the edge of downtown KL, at about two A.M. Sha first drives past the residence of Anwar Ibrahim himself, before we park on the edge of the main commercial area and walk down streets lined with Italian bistros, Irish pubs and even a Starbucks to an unassuming outdoor restaurant for tea. Honking, flag-waving motorcades drive by intermittently: first, a screaming, banner-bedecked horde of government supporters, then one for the opposition party of Anwar's wife, then one for the Islamic opposition party, and another for the government. The feel is markedly similar to that of a city whose team has just won the big game, particularly because there appears to be little direct confrontation (or even much animosity) between the motorcades flying different-coloured flags. It's as if they're celebrating the very existence of an election.

Sha speaks about all the stuff that Dr. M has done to improve Malaysia's self-esteem—citing, for example, the government's recent financing of the efforts of the first Malaysians to reach the summit of Mount Everest. "He shows us that Malaysians can do anything if we put our minds to it," Sha concludes. All of this is encapsulated in the slogan *Malaysia boleh* ("Malaysia can"), which began as a sporting slogan but has been adopted by Dr. M and—judging by its prevalence on everything from cocoa ads to waiters' aprons to bumper stickers—the whole nation.

I am reminded of something I saw on TV—an ad for Malaysia Airlines featuring images of Malaysian firsts, shot in gorgeous black and white and set to the triumphal strains of Frank Sinatra's "My Way." At the end, a voice-over intones: "We can do it, if we do it our way." This twist on *Malaysia boleh* comes even closer to summarizing Dr. M's style, which was in abundant evidence during the country's recent economic crisis and is, I suspect, part of the reason why many in the West are eager to find reasons why Malaysia, in fact, can't.

PRIOR TO THE CRISIS, IT WAS HARD to find a dissenting voice on the subject of Dr. M's leadership and the country's potential. *The New York Times*, for example, met the announcement of the MSC's launch in early 1997 with the headline, "Silicon Valley of the East? Try Malaysia." That July, Thailand's currency collapsed, setting off a regional crisis that saw Malaysia's ringgit lose sixty percent of its value and the KL stock exchange lose at least $100 billion within a couple of months. As exchanges continued to sink, Globalization saw fit to send one of its acronym minions—the IMF—to save east Asia. It was at this point that Dr. M committed his ultimate heresy: He refused an IMF bailout.

Instead, he continued to spend freely—on such suddenly superfluous "pet projects" as the MSC—and implemented currency controls, fixing the ringgit against the U.S. dollar to stabilize it. All of these were the near-exact opposites of austerity recommendations made by both the IMF and Dr. M's own deputy and finance minister, Anwar Ibrahim. In short, Dr. M did it his way. And then the press turned on the MSC. In one high-profile example, *Business Week*'s March 1999 cover story (entitled "High-Tech Folly") revealed that foreign investment was progressing more slowly than he had originally hoped (though a near economic meltdown will do that sort of thing).

But by the fourth quarter of 1999—and this must've really irked the so-called experts—Dr. M's economy recovered. It grew by about ten percent. And then everyone from a team of American congressional staffers on a trade mission to the IMF was forced to admit—grudgingly—that they were wrong. David Marinelli, president of Lucent's Malaysian operations, put it to me more bluntly: "There's a lot of people eating crow right now, because of what they thought they knew better than anybody else."

The results of the Malaysian election appeared in the papers on Wednesday morning, accompanied by lurid photos of street violence. Dr. M maintained his two-thirds majority, without incident; the photos were from the WTO protests in Seattle. The demonstrators in Seattle were a mixed bunch, to be sure, but from the distant vantage point of KL, it looked like they were demanding that Globalization relax its dogma a little—that its acronym avatars take note of their interests, as well as those of the corporate world. Not that different, really, from the demands of an IMF-defying, currency-controlling prime minister I know.

PINK HOUSES

NOT LONG AFTER THE ELECTIONS, Sha and I take a drive to Putrajaya, Cyberjaya's planned twin city and the seat of the wired, paperless government of Malaysia's near future. (This wired-administration plan actually predates the MSC project, and is, in some respects, its genesis.) Putrajaya lies due east of it, across a few miles of an oil plantation that, if all goes according to plan, will be filled in by both cities' growth by 2020. Even more so than Cyberjaya, Putrajaya rises out of the palms like a mirage. You're driving through wilderness, you round a bend, and suddenly there's a whole *city*—or at least the nucleus of one—on the horizon. Putrajaya is centred around a broad, half-completed promenade, which runs from the foot of the main government building to, well, nothing at the moment. The government building itself—which Dr. M and his staff have already moved into—is an enormous stone edifice, several blocks wide, that looks like a classical-style capital building, except for its crown of Arabic onion domes. To one side of the promenade is Masjid Putrajaya, a huge mosque with an intricately detailed dome in pink marble. Spreading out from this core is a mishmash of bland office buildings and neat rows of townhouses, plus the prime minister's palatial residence. The whole place—grand, but not quite ostentatiously so—feels triumphal. And, by comparison to Cyberjaya, *bustling*. The parking lots are full, bureaucrats flit to and fro, a handful of faithful lie prostrate inside the mosque. This is not a dream, not some far-fetched vision. It exists.

Sha runs ahead of me, up the steps to the immense front doors of the main building, and then turns back with a big grin. "Next time you come to Malaysia," he calls, "you come visit me here. This will be my office!" Sha has mused aloud on occasion about wanting to become prime minister—among other plans—and he talks about these ambitions with the kind of forthright optimism usually heard in movies starring Jimmy Stewart. I half-expect him to start his next sentence, "Gee whiz…"

And heck, why not, Sha? That is, after all, what Dr. M is getting at, sponsoring treks up Everest and all. Everywhere you go in the KL area, you run into giddy assertions that life is being lived in a future tense in which anything is possible. Here, you can find content providers partnering with the web's biggest names; e-commerce start-ups ready to take on the industry's heavyweights, or list on Nasdaq, or both; and design shops launched by heretofore cyberilliterates. At present, 302 companies have been granted MSC status; of those, fifty-nine percent are Malaysian-owned, and another twenty-one percent are joint ventures between Malaysian and international companies (which is to say, the MSC is not just a free-trade zone for multinationals). And they're almost universally infused with the same kind of frank optimism that Sha has for his political ambitions. I mean, there is a URL prominently featured on the sign in front of KL's national mosque.

But it is not, I should add, *starry-eyed* optimism. The heads of MSC start-ups recognize that Malaysia has almost no venture-capital industry to speak of, and what little exists is in the hands of people who, in one MSC entrepreneur's words, "don't understand the nature of risk and return, and don't understand the nature of the internet industries." (The government, in an effort to address this very problem, has started its own venture-capital company in recent months.) And they recognize as well that serious sales and marketing occur elsewhere; a number have offices in Singapore, Hong Kong and Silicon Valley—the same places they go looking for venture capital—for marketing purposes, and use their MSC offices primarily for R&D.

But they also recognize, in the end, that they are part of something very big, and very exciting. Very few, though, are exhilarated enough to move to a "city" with few ancillary services (restaurants, for example) that requires a commute to the boonies. But in time, who knows—some of them may actually relocate their offices to Cyberjaya.

SPACE ODDITY

"YOU CANNOT DO ANYTHING," Dr. Mohamed Arif Nun is telling me, "if you do not have a house." We are seated in a large, somewhat barren office at the Multimedia Development Corporation in Cyberjaya—the office Dr. M uses when he has work here—and Dr. Arif, the MDC's senior vice-president, is patiently explaining why the Malaysian government is sponsoring so much construction out here in the sticks.

He is a soft-spoken, congenial man, fifty-four years old, and he's more comfortable talking about things like internet IPOs, smart cards and convergence than any other fifty-four-year-old I've ever met.

He also has a quick wit; earlier in our discussion of the pressing need to get companies to relocate to Cyberjaya, he said, "It's hard to build a collaborative, new-age, third-millennium cybercommunity if someone is in the Rocky Mountains and someone else is on a Pacific island." He exaggerates, but the fact remains that MSC companies are currently scattered all over the KL area. All of them are required to relocate to Cyberjaya (or one of KL's four other "designated cyber-cities," a term that includes the Petronas Twin Towers and three tech parks in the suburbs) by July. If they don't, they risk losing their MSC status, along with its concomitant tax holidays and duty exemptions. I doubt, though, that the MDC will boot anyone out of the MSC community anytime soon. I have to add here that "only" twenty-five companies have set up shop in Cyberjaya to date, but the city is less than a year old, and the majority of its existing office space has been leased. Blame it on the sense of utter urgency the MSC has created around itself; this very office complex we're sitting in, after all, was built in ninety days by crews working around the clock.

"If you want to send the space shuttle or a satellite into orbit," Dr. Arif says, on the topic of creating just such a feeling, "the first phase is always difficult—which is giving this rocket the eighteen-thou-sand-mile-per-hour boost, with the three engines that drop off, and finally reaching about eighteen- to twenty-thousand miles per hour to actually—what do you call it?—to reach escape velocity."

But the Malaysian government faces another, trickier problem, as Dr. Arif concedes: It needs to find the right astronauts, the kind of people who will not just fly to outer space (Cyberjaya), but will encourage others to go, as well—ideally dragging along their favourite restaurants and shops, and turning this development project into a full-blown city.

"What you need to shortcut this process"—the speaker now is Derrik Khoo, the CEO of an e-commerce company called Go2020.com—"is a quick success story, and it has to be local. Everyone knows that. You can have Hewlett-Packard there, IBM, everything, and people will still ask that one question: 'Name me one local company you've produced.'" For the record, after a long lunch with Derrik Khoo, I'm ready to name Go2020.com.

First, there's the company itself, which, like any good e-commerce start-up, is working on a product that is quite difficult, at first, to make any sense of. So, like any good explanation of what an e-commerce company does, this one starts with an analogy: Go2020.com's product aims to be to e-commerce portals what Microsoft is to software—that is, virtually shrink-wrapped, off-the-online-shelf e-commerce applications for all those businesses that may not need ultracustomized, one-of-a-kind e-commerce solutions, but merely need *an* e-commerce solution. Got it? No? Well, never mind, because I guarantee Khoo could sell you on it, even if you didn't have a clue what it was.

Sit with him in a restaurant in a KL suburb, watch him gesture with his pizza slice and lapse occasionally into a truly goofy Italian accent, spinning yarns, and you'll find yourself mesmerized, invigorated, elated. You will want to be on his team. You will probably want to give him money. He knows how to talk almost exclusively in analogy. He has that info-age star quality the MSC so desperately needs. In short, he's a kick-ass high-tech entrepreneur. Hell, he even looks a bit like a Chinese Bill Gates, with his round, wire-rimmed glasses, crooked haircut and slight, boyish features. And both his chief operating officer and his chief technology officer were poached from Microsoft's Singapore office. The latter, Cheah Kai Kooi, who has joined us for lunch, is soft-spoken and a bit older than Khoo, and he talks with the quiet, bemused intensity of a true inventor about the engine he's built. When he's done, he and Khoo go off, riffing, analogizing and explaining why Kai would want to leave Microsoft—and why Khoo would want to leave a cushy gig as an editor at KL's largest newspaper—to join Dr. M in his "ridiculous, expensive, preposterous idea." Take a listen:

> KAI: Imagine you are driving your Porsche, going to Beverly Hills. What is it? It's not outstanding, right?
> KHOO: Yeah, it's nothing. Depends on the Porsche. Is it the latest one?
> KAI: But even the latest one, it doesn't matter. It's not outstanding. Now, imagine you drive the same car in … Bombay. *Wow!*
> KHOO: Everybody will be wondering how come he's having a drive like that in Bombay. So, yeah, you'll be *outstanding*, all right [*laughs*].

KAI: So that's the way. You go into a place where everyone can buy it, you will be average, whereas here, the MSC is new and you're the only thing here. It's from the ground up.

KHOO: [*turns to me*] But, of course, the example that Kai gives you is a double-edged sword. There's a reason why there's no other Porsche in Bombay. Pretty soon, if you have the only Porsche in Bombay, you'll be *not having* a Porsche anymore in Bombay. It will probably not be well maintained; the roads are probably not conducive. That's why the government built the MSC.

I might bicker with the choice of cities—you're probably more likely to see a Porsche in Bollywood than, say, Des Moines—but I don't disagree with the sentiment. And it occurs to me, talking to the Go2020.com team, that even if their company turns out to be something less than a Porsche, even if it doesn't succeed on a massive enough scale to give Cyberjaya the jolt it needs, even if it fails completely—regardless, *something* will come along to provide that jolt. I'm feeling pretty sure, that is, that the MSC has reached escape velocity.

"You can make fun of anything you want about the *Malaysia boleh*, the Malaysia can-do," Khoo tells me. "But actually, when you're trying to sell a dream, when you want people to believe in themselves, you have to also believe in the *Malaysia boleh*."

ROAD TO NOWHERE (REPRISE)

THERE ARE CERTAIN NUANCES in the phrase *Malaysia boleh* that get lost in the translation.

To wit: I had moved to a hideous corporate hotel called the Hotel Equatorial in the KL suburbs, hoping that it would make the city more accessible. The staff at the hotel insisted that I could not simply call a cab to take me back to Cyberjaya for an interview I had scheduled; that I must use its limousine service. I reluctantly agreed, not expecting they meant an *actual* limo. They had. It had a slight stretch to it, and a bar.

Somewhere around halfway there, it began to rain—serious rain, falling in great steaming sheets—and we came to a half-completed overpass, where a line of neon traffic barriers blocked our lane. There was a sign on one of the barriers in Bahasa Malaysia with a big arrow: LENCONGAN. The driver turned around the way we came, saying simply, "Road close." We backtracked to another highway and drove on it for a while. Took an exit. Stopped to ask directions. Drove the way we'd just come. Stopped again for directions. I started to ask questions like "Do you know where we're going?" and "You don't know where we're going, do you?" He told me the roads had "change."

Finally, around the time I'd missed my meeting completely, he stopped one last time for directions and then drove off again. The interview had been set up by my main contact at the MDC, who had doubtless waited for me in vain, who was doubtless disgusted by now with my irresponsible, if not downright unprofessional, behaviour. Would he now be so angry with me that I'd never secure another interview in this town again? We came once more to the overpass with the barriers and the sign with the big arrow. LENCONGAN had, in fact, meant "detour." We switched lanes, drove on, and Cyberjaya came into view just beyond the overpass.

My contact, Raslan Sharif, greeted me in the MSC headquarters lobby with a broad, knowing grin. "What happened?" he said warmly. I explained; he laughed heartily. He said he'd figured I was "caught in the rain." He would reschedule the interview. He would drive me back to my hotel, if I didn't mind—*if I didn't mind*—waiting a little while. I went to the Cyberview Lodge bar and ordered a drink. *Malaysia*, I thought to myself, *boleh*. And then, for reasons unclear at the time, I laughed till my sides hurt.

There are two ways to interpret this little anecdote. The first is to think, Hell, if you can't even find a cab—a *limo*—capable of getting you to your meetings on time, these people are *dreaming* if they think the high-powered, speed-of-light-paced digital world is going to set up shop here. No way a place like this is going to become the next Silicon Valley—and then laugh at the MSC's folly.

The way I see it, though, I was laughing at myself—for getting so worked up, for thinking that my Malaysian contacts would react the same way my North American ones would. For thinking, in short, that Malaysians were busy building the next Silicon Valley. They're not. They're busy building the first Multimedia Super Corridor.

A MISUNDERSTOOD SUBCULTURE, A VEGAS RESORT AND LOTS OF BLACK T-SHIRTS, LAPTOPS AND BOOZE

Shift, November 2001

My old university friend Joey DeVilla, a software developer, first suggested I tag along with him to DEF CON. "Hacker bacchanal in Vegas"—maybe the easiest pitch I ever sold. The reality was more complex, a whiplash mix of laughably straightforward assignment (attend conference, describe what happens) and tricky execution (every possible subject, for example, warier of journalists than any I'd ever again encounter).

For all of that, the mix of wild Vegas fun and crazily high-level conversation made this an unexpectedly educational vacation of an assignment. I remember coming back to my cheap hotel (the Holiday Inn on the Strip) one evening at like four in the morning, having spent the day listening to the world's foremost security expert (Bruce Scheier) give a mind-blowing lecture and then drinking at the Cult of the Dead Cow's hospitality suite. A note-perfect Prince cover band was playing in the dingy little Holiday Inn cocktail lounge. The costumes, the voice, the mannerisms—flawless. What else could I do? I stayed up for one more drink.

I would never again mimic David Foster Wallace as fully as I did in this piece. I don't regret it and it suited the material, but it's a dangerous path. You'll never beat the teacher, and you may prevent yourself from acquiring other tricks of your own.

THE GUY UP ON THE PODIUM identifies himself only by the handle "Shatter." He's half-hidden behind a laptop and cables and networking gear, but I can still see his black T-shirt, his goatee, his long hair pulled back in a ponytail to reveal shaved sides—a fairly common post-punk, non-conformist hacker look. Shatter is walking a full-house crowd through a PowerPoint presentation—his topic is, roughly, Hacking 101—and he's doing it with an expert mix of straightforward detail and humorous anecdote. He tells one about how, back in 1993 or so, he used to trick script kiddies—beginner hackers who use serious hackers' programs (or scripts) to carry out the digital equivalent of petty vandalism—into handing him complete control of their computers. The punchline, wherein Shatter's script goes about erasing the script kiddies' home directory, just *kills*.

The setting for this presentation is the Zeus Ballroom of the Alexis Park Resort in Las Vegas, Nevada, day one of DEF CON 9 (DC9)—this year's incarnation of DEF CON and an A-list event on the social calendars of hackerdom's best and brightest—and gags at the expense of script kiddies are right up there with *South Park* references in the humour hierarchy.

Shatter gets a few more big laughs with a quick survey of popular hacker misconceptions—like getting caught hacking government sites will be rewarded with a government job—and then moves toward wrapping up. A PowerPoint slide pops up on the big screen to his left, headed "Hang Over Remedies" [sic]. And Shatter says: "Alcohol and DEF CON go hand in hand." It's a revelation that surprises exactly no one, even in this room full of newbies.

DEF CON: THE LARGEST UNDERGROUND INTERNET SECURITY GATHERING ON THE PLANET—OFFICIAL DEF CON WEBSITE [DEFCON.ORG]

THE BASICS: EVERY SUMMER SINCE 1993, the elite of the hacker world has gathered in Las Vegas for a conference-cum-beerbash called DEF CON. The 2001 version mapped out like this: About sixty seminars in three streams of expertise ("Newbie," "General" and "Uber Haxor"), with about seventy speakers in all. One of said speakers (a Russian coder named Dmitry Sklyarov from the Moscow software company ElcomSoft) arrested, post-conference, under the Digital Millennium Copyright Act. One rousing rallying cry for a political movement by the name of hacktivism. Three evening rounds of "Haxor Jeopardy," with attendant heavy drinking and clothing removal by crowd favourite "Vinyl Vanna," and an equal number of nights of live DJs and the odd industrial band. A host of other organized events, including the ever-popular "Spot the Fed" competition and the Black & White Ball (this last involving fairly elaborate costumes in some cases). One hotel payphone forcibly removed from the wall and deposited in a toilet. At least one steamy, late-night makeout session between a transsexual and a sloshed, unsuspecting attendee in an Alexis Park hot tub. One poolside consumption of a simply enormous cockroach, with the consumer—a solidly built, good-natured sort named Bill—claiming a (U.S.)$181 reward cobbled together from onlookers' donations. An estimated food-and-drink tab, at the Alexis Park alone, of (U.S.)$70,000—the resort's highest-grossing weekend of the year, exceeding even the Royal Air Force's annual bash. Average temperature in the low forties, Celsius, during the day, high twenties in the evenings. Humidity seemingly less than zero percent. Total attendance was around 5,000. Estimated male-female ratio was perhaps 6:1.

One high-profile speaker's summation (to Reuters news service): "DEF CON is a cross between a Star Trek convention and a Ramones concert." My own: Spring Break for Hackers.

The first DEF CON, by contrast, was a modest affair, attracting about 100 or so friends and acquaintances of the organizer, Dark Tangent (a.k.a. Jeff Moss), to the Sands Hotel and Casino in July 1993. In subsequent years, the conference got progressively bigger and rowdier, and it changed hotels annually—often because the previous year's host had decided it could live without another round of the DEF CON crowd's unique breed of fun—until it found its semi-permanent home at the Alexis Park in 1999. By this time, the "underground" nature of the event had eroded considerably: The mainstream media and legitimate computer security firms were attending in greater numbers, and even the once-edgy "Spot the Fed" game became less about outing the enemy and more a good-natured joke between peers. In fact, U.S. federal officials from an alphabet soup of security and justice-related agencies have just come right out and sat on panels at the last two DEF CONs, and even the most clueless of wire-service stringers now manage to find one or two "undercover" agents to add colour to their stories.

These days, DEF CON's focus is as much social as professional, with the serious security-industry stuff relegated to a pre-DEF CON event: a (U.S.) $1,000-per-head, business-minded affair called the Black Hat Briefings, which occurs at a separate Vegas hotel in the days leading up to DEF CON. Or as Dark Tangent, who organizes both conferences, put it at a casual press conference on Friday afternoon, "Black Hat was the university and this is the frat party."

So let's check out the frat party. It's mid-afternoon in searingly hot July. Directly below the press area is the Parthenon Ballroom, a gymnasium-sized space that serves as DC9's central meeting area. It's got a bar in the middle, tables scattered around, and its walls—like the walls of nearly every hallway in the Alexis Park all weekend—are lined with hackers sitting in ones and twos on the carpet, tethered there by cords running into electrical and phone jacks, doing I-couldn't-possibly-tell-you on the laptops in front of them. The only thing I know for sure is that not one screen I pass, the entire weekend, is running a Microsoft operating system.

The Parthenon has two side wings. One is crammed with conference tables, which are draped in black linen, piled high with laptops and other digital-communications paraphernalia and surrounded by dense gaggles of conference-goers. They stare intently at various screens, some occasionally tapping out a few lines, making very little noise collectively beyond the clacking of keys and the odd explosion of cheering from a table full of hackers who have just done something spectacular on DC9's computer network. This is the Capture the Flag competition, DEF CON's most prestigious contest, in which teams of hackers attempt to do things on the network set up for the conference that are far too elaborate to get into here. The other side wing is the vendor area, which is an even mix of tees, vintage electronic equipment and strangely titled books. The main lobby and various lounge areas are given over to tables full of drinking, laptop-tapping and/or socializing hackers. And the nexus of all of this is the apron around the main pool.

So I'm out by the pool now. It's mid-afternoon on Saturday, cloudless, the desert heat steady and dry like a giant oven's been left open a few feet in front of me. DC9 is in full swing and the vibe is collegial. If you're here, you must be one of us (or wish you were), seems to be the assumption. Plus there's a disproportionate number of people here who are not used to being surrounded almost exclusively by people who truly get them, I'm guessing, which means everyone's in a pretty friendly mood. But then there's no shortage of stereotypes about hackers in this world and I'd rather not add to the pile. So view all this as a series of Polaroids taken from the pool deck at DC9 and draw your own conclusions.

First: At any given moment, the ratio of black T-shirts to all other colours and styles of shirts stands at about 1:1, spiking to 2:1 when a full posse of head-to-toe-black-clad hackers struts by. (This is the most visible evidence of the robust cultural cross-pollination between hackerdom and the goth/industrial music scene.) Every so often, a "greybeard"—an old-school Unix hacker with requisite flowing whiskers—appears. Here's one, for example, in aqua swim trunks and a black tee emblazoned with a Microsoft dis, his beard dense and grey and flowing to mid-chest, his head nearly bald. A middle-aged fellow passes, decked out in business ultra-casual (company tee, denim walking shorts, white running shoes) and sporting a Republican haircut and a paunch. Then a kid, early twenties, with a pink mohawk, skater shoes and a laptop tucked under his arm in lieu of a board. A black man—something of a rarity, East and South Asians (in that order) being the most numerous visible minorities at DC9—strolls by, spare-tired, floral-shirted and clutching a freshly purchased black T-shirt reading "SUSPECT." You could do a detailed anthropology of this scene entirely based on things written on people's T-shirts. *Don't get caught. Go away or I will replace you with a very small shell script. I will not instigate revolution.* And so on.

At a guess, I'd say the DC9 crowd was equally divided between computer-security professionals and people whose day job is something else. Most hackers work full-time or on contract in the computer industry, or are students. A small sliver left of DC9 attendees are media, gawkers, government officials, and friends of hackers—who are also often employed in the computer industry but are not themselves hackers and who are here for the party. There are no official sponsors of DC9, but I'd give the unofficial nod to Red Bull energy drink: It's on sale all over the place, including that logoed cart over on the far side of the pool, which is hawking it mixed with vodka and doing a brisk trade. The women here almost universally exude a self-assuredness born, I bet, of having spent most of their professional lives as the only woman in the room.

And at one point, a staggeringly curvaceous young woman appears on the far side of the pool. The largest single item of clothing she's wearing is her DC9 ID badge. She's sitting on a lounge chair in a position that reveals too much even for most cable channels, and pretty soon nearly all of the people seated or standing on this side of the pool are stealing increasingly less and less surreptitious glances at her, and finally someone cracks and goes over and asks for a photo. A small dam seems to break and an impromptu lineup forms, a row of digital camera-wielding twentysomething hackers waiting to have their photo taken with this woman, and she deals with it all in a good-natured, semi-professional sort of way that makes me think she knew exactly what would happen when she wore that particular swimsuit to the pool at DC9.

All right. Let's say I'm sitting in the same spot eight, ten hours later. Coming up on midnight, say. I take another set of Polaroids. What then? First: Infinitely fewer middle-aged corporate types. Second: Way more leather and PVC. This is especially true on Saturday night, the night of the Black & White Ball, when the goth contingent in particular goes all out. One woman, for example, wears a black minidress and Hollywood-quality black angel wings. A couple in bondage gear make out conspicuously in the main lobby for some time, and he pulls up her dress to reveal a G-string and spanks her loudly for a full minute or so. Down by the far pool, there is a keg party and fire breathers. One guy cruises the whole scene in a PVC tank top and a World War II-era gas mask. Also: The pools are full of people, most fully dressed, some hardly dressed at all. The odd overindulger is wiped out on the grass or in a lounge chair and there's plenty of other collegiate hijinks in evidence (staggering drunks, loud yells of *Whooo!*, off-key singing, sliding down banisters, that sort of thing). Many of the hotel's larger suites—two-storey, multi-room, full-kitchened affairs—are given over to private, sometimes even invite-only parties hosted by the more prominent or ambitious hacker groups.

And, perhaps most remarkably, there are still tables full of laptops and small knots of hackers grouped around them, still doing their thing. If this weren't DEF CON, I'd say they were still working, but that's not quite right; it only tells half of the story. To tell the rest—much as I'm worried that any attempt at generalizing, no matter how empathetic, may bring down upon me the ire of people capable of destroying my email account, my credit rating, my entire hard drive—still, to tell the rest, I've got to go into what hackers are, exactly.

FRIDAY, JULY 13, 11:00-11:50: BRUCE SCHNEIER ANSWERS QUESTIONS

THE ROOF OF THE ALEXIS PARK Resort's main building was crowned, for the duration of DC9, by an enormous white tent that served as the auditorium for the conference's top-tier speakers, and it was mercilessly hot. There were enormous ducts coming off its sides at various points—some kind of cooling system—but these were clearly inadequate, and these, combined with the greenhouse effect created by the tent itself, meant the interior of the place was maybe a degree or three cooler than the outside air temperature but infinitely more humid. The guy sitting next to me on Friday morning had this blue, semi-transparent, plastic horseshoe apparatus around his neck called the "Personal Cooling System 2.0," and though my first instinct was to mock his Home Shopping Network taste, I was soon envious.

Enter into this steam bath one Bruce Schneier—author of the authoritative professional-hacker text *Applied Cryptography* and widely regarded as one of the planet's leading experts on computer security—a slight, thin man in beard and teal-coloured golf shirt, his hair worn long despite a dramatically receding hairline, and his self-confidence, at this event, strong enough to be almost tangible. Schneier strutted the DC9 stage like a geekified Mick Jagger, tossing off one-liners and speaking with casual authority on everything from the future of online music to the probability of communications-network disasters on a global level. (Very likely, by the sound of it.) Halfway through Schneier's speech, it was standing-room-only in the tent, despite the heat, and the mood was palpably reverent. Someone asked Schneier what his background was. He paused for effect, turned, inspected, and said, "It appears to be some kind of black velvet curtain." *Huge laugh.* Ask around, he told the poor newbie who asked the question. You'll figure it out.

Point taken: The hacker world is a robust enough subculture to produce its own starmaking system, and it's a world where an expert mathematician can become a full-blown rockstar. So hang *hacker* up there on the wall of pop-cultural archetypes next to *hippie* and *punk* and *preppie*, I guess. But note, also, that it is an elastic archetype by modern pop culture standards, limited not just to teens and twenty-somethings of a particular fashion sense and worldview, but to anyone from the middle-aged crypto expert to the teenage goth. The only real qualification, it would seem, is that you are the type of person who looks at any given machine and asks not *What does it do?*, but rather *How does it work? And, moreover, How else could it work? What else could it do?* And you must find these kinds of questions endlessly fascinating.

Consider: One of the private Saturday night parties in a large Alexis Park suite—one of the more exclusive ones, by the look of it, with printed invitations and a bouncer and everything—was called Caesar's Challenge. The front room of the Caesar's Challenge suite had been converted into a kind of lounge. There was a DJ and a cocktail bar, and the walls were draped in metallic wrapping paper and white Christmas lights. Most of the action, though, was in the back room of the suite, which was crammed full of people and steaming hot, even after midnight. The walls of this back room were plastered in butcher paper, which were in turn plastered with diagrams, equations and explanatory text. This was the "challenge" part of Caesar's Challenge. There was a handout with some hacker-related brainteasers on it and knots of people—concentrating intently on the task, even the ones who were teetering drunk—were trying to solve them. After midnight on the biggest party night of the biggest bash of the hacker's year, your true hacker is still more than happy to kick around computer-science problems.

Out by the pool, others were doubtless doing similar mental calisthenics in front of their laptops: trading programs, writing code, whatever. One attendee I know spent his downtime at DC9 at the Alexis Park's bar, using a "sniffer" program to pluck user IDs and passwords off the conference's wireless network. On Saturday afternoon, a long-haired Texan explained to me how he could "hack" the DC9 ID badge, which consisted of a sealed plastic sleeve full of coloured liquid, with cutout letters and symbols floating in the liquid. His solution—which involved inkjet refills and a syringe, I think—is beside the point. The really amazing thing is that this Texan fellow's natural train of thought, as he looked down at the badge, rubbing at it to make the letters line up, was: *How could I hack it? Could I duplicate it? Could I change its colour so people thought I was a speaker, a security guy, a press guy?*

SATURDAY, JULY 14, 15:00-15:50: MEET THE FED PANEL

IF YOU'RE NOT A HACKER, you might still recognize the name Bruce Schneier. That could be because he wrote the appendix to Neal

Stephenson's *Cryptonomicon*, an epic novel that does for the hacker world roughly what *Moby Dick* did for whaling. And if you've read *Cryptonomicon*, it would likely occur to you that if Schneier had been the author of the leading English-language cryptography text prior to 1945, he would not have spent his time speaking to rooms full of predominantly marginalized young men and running a modest private consulting firm. No, he would have been an invaluable expert adviser, high up in the military-political establishment, working on projects that would determine the fate of the industrialized world. And so, I would argue, would a great many of the attendees of DC9.

During World War II—and even into the fifties, as the U.S. government assembled its Cold War national security state—hackers (by other names) were a vital part of the military-industrial complex. In an ironic twist, the explosion of digital technology over the few decades since then has been accompanied by the relentless marginalization of the people who best understand it. As the cryptographer (a highly specialized species of the harmless mathematician genus) mutated into the hacker (a dangerously antisocial breed of computer-obsessed deviant), the powers-that-be turned their best friends into their worst enemies. World War II-era hackers were key advisors to presidents and the Allied High Command; today's hackers have been treated like criminals since at least the early 1980s, when *War Games* hit the Cineplex. In fact, if there is anything uniting hackerdom beyond a passion for meddling with electronic machines, it's a near-universal sense that they are fundamentally alienated from, and misunderstood by, mainstream society. Never mind that 5,000 of them are bopping around Vegas like they own the place: These are people who have been overlooked for their technical skills, abused for their interpersonal skills and outlawed for their curiosity.

A significant swath of hackerdom has come to embrace its marginalization. This is the root of the traditional DEF CON "Spot the Fed" game, which rests on the premise that government representatives want only to spy on and imprison the kind of people who attend DEF CON. Though initially an exercise in group paranoia, "Spot the Fed" is now one of DEF CON's most beloved contests, permitting any attendee to publicly "out" Feds in the audience. If you're right in your accusation—the Feds generally admit their affiliation—then you get to strut around the conference for the rest of the weekend in a T-shirt reading "I spotted the Fed"; the Fed, meanwhile, gets an "I am the Fed" tee. No T-shirts were handed out at Saturday afternoon's "Meet the Fed" panel, though, when a handful of high-ranking U.S. government officials—including at least one from the National Security Agency (NSA)—took the stage under the tent.

The "Meet the Fed" panel is the Establishment's attempt at explaining its role in the hacker world. It's a more recent tradition than just spotting Feds, and it says a lot more about the current state of flux in the relationship between hackers and the Feds. Until last year, the Feds' role was simple: They were Eliot Ness and hackerdom was Capone. But DEF CON 8 was a sort of watershed in Fed-hacker relations. Not only did Feds take the podium for the first time, but an NSA rep came right out and asked for the hackers' help, and a Department of Defense official all but admitted that his department couldn't keep its information secure. At DC9, according to one British newspaper report, the same Defense Department official was set to make job offers. I have to wonder if he's too late.

It's not that DC9 is crawling with anarchistic, tear-down-the-system types—though there were likely a few in attendance, curiosity is a far more universal motivation than anarchy in these circles. But even as older hackers drift toward paying gigs and ethical stands, few of them are likely to be convinced that the U.S. government, for example, is their kind of people. Especially not, it warrants mention, when the big news coming out of DC9 was not anything said at the "Meet the Fed" Panel but rather the FBI's arrest, at the behest of Adobe Systems Inc., of Dmitry Sklyarov.

On Sunday afternoon, Sklyarov, a Russian programmer, gave a seminar—the final "Uber Haxor" seminar of DC9—on how to get around the copyright protection on some of Adobe's products. I stumbled into it on a mission to attend at least one "Uber Haxor" event, and whatever was being presented in that conference room was way over my head and delivered in a thick East European accent to boot. Still, I feel pretty confident in telling you that the scene I witnessed—a darkened room, rows of quiet, tired and/or hungover-looking DC9 attendees looking on, PowerPoint slides on an overhead, someone speaking in a flat, heavily accented drone—did not reek of intrigue, of danger, of federal offences being wantonly violated. The next day, though, Sklyarov was arrested for violating Adobe's copyright, or, more accurately, for explaining how someone could. This kind of casual censorship speaks louder in the halls of hackerdom than a thousand warm, friendly Feds talking from the podium. Within days of Sklyarov's arrest, two websites—freesklyarov.org and boycottadobe.com—had been set up in his defence, and there have since been protests at the San Francisco courthouse where he was arraigned. And hackers, it is revealed, can be plenty political if you push them hard enough.

SATURDAY, JULY 14, 14:00-14:50: cDc HACKTIVISM PANEL

IT'S OPEN-SWEATING WEATHER inside the tent by the time the hacktivism panel is introduced on Saturday afternoon, and so a significant number of audience members are fanning themselves with DC9's tabloid-sized programs, which creates an odd oscillating effect. The keynote speaker is one Dr. Patrick Ball, Deputy Director of the American Association for the Advancement of Science's Science and Human Rights Data Center.

Dr. Ball is no hacker: He gathers data on flagrant human-rights abuses—like death-squad atrocities and genocide-level extermination campaigns—in the interest of helping bring the abusers to justice. He is here as the guest of the Cult of the Dead Cow (cDc)—one of the world's longest standing and most prominent hacker groups—which has been doing some sort of major presentation at DEF CON every year since 1997. Their launches, in 1998 and 1999, of the first and second versions of BackOrifice (a now-legendary piece of software that turned Microsoft's feeble firewall, BackOffice, into a swinging door) were peak points at both conventions. Indeed the cDc's presentations can be seen, I suppose, as the bellwether of DEF CON, the state of the union address from some of hackerdom's most senior officials. Or something like that. This, anyway, might explain why the crowd nearly overflows the white rooftop tent at the Alexis Park on Saturday afternoon, despite the ever-more-staggeringly hot interior.

Ball opens his talk by referencing the Quaker doctrine of speaking truth to power, and it's instantly clear that this is no acronym-drenched hacker lecture. Ball is fairly short and stout, bespectacled and bearded, more dwarfish than statesmanlike in his bearing, but he's got a far more classically political speaking style than anyone else I've seen here, punching just the right words for emphasis and varying his cadence to create momentum. And it's working. Ball's work involves turning reams of eyewitness testimony into hard stats on systemic human-rights abuses, such as tracing hundreds of murders in El Salvador back to high-ranking military officials. His speech thus focuses on visceral, emotional topics. As he's talking, a sudden, ferocious wind kicks up outside, and the tent's walls and flaps are undulating and rolling in time with the programs that everyone is flapping, and the whole metal frame of the tent is shucking and jiving in the breeze, punctuating Ball's points with loud, wrathful groans, and Ball is working himself up into a righteous anger. This is not talk of obscure technical data or mocking Microsoft; this is a rallying cry. This is about changing the world!

When Ball finishes, there's a great, heartfelt roar from the crowd—greater even than anything hacker hero Bruce Schneier got—and a few people even leap to their feet, clapping madly.

So how did one of DC9's most eagerly anticipated presentations become a forum for a mostly tech-free political oratory? The main reason is Peekabooty, a piece of software currently in development by the Hacktivismo group—a "special operations group" sponsored by the cDc—which promises to allow people living in countries where the internet is censored to gain unfettered access by networking their computers to activists and sympathizers in censorship-free places. How it does this is, naturally, quite complicated—so much so, in fact, that the Hacktivismo group backed off on its initial plan to launch the software at DC9 because it wasn't ready yet. There was grumbling in some corners that this meant the cDc had lost its edge, which to my mind misses the point. Namely, as Hacktivismo founder and cDc foreign minister Oxblood Ruffin explained it to me over beers at the bar of the Hard Rock Hotel after the panel, that Hacktivismo represents a new level for him and his colleagues, a decisive step onto a much larger and far more serious playing field. There was a genuine edge to Oxblood's voice when he told me this—an interesting mix of fear and righteous excitement. Menacing hacker handle aside, Oxblood is older and more contemplative than most hackers, with a definite intellectual bent, and so now that he finds himself not just talking up the injustice of the Chinese government or the oppressive regimes of various Middle Eastern states but actually working toward releasing something into the world that would challenge them directly, he is not naive about the import of it.

There had been similarly charged talk at the cDc suite at the Alexis Park the previous night. I had spent the better part of the evening talking about China's occupation of Tibet—and related information-flow and censorship issues—with a few of the people from the hacktivism panel and some of the cDc's more politically minded members. If the rest of DC9 could be seen as a sort of hacker frat party, then the cDc suite came across on this night like a kind of hacker salon: Conversations tended to veer toward political topics, some people in attendance were drinking microbrews, and there was a recent issue of *Foreign Affairs* sitting on the back of the suite's toilet. At one point, one of the panelists talked about all the "latent power" he'd seen on display at DEF CON, about the enormous potential that DC9's attendees represented. It was something I'd nearly forgotten, a fact washed away in waves of cheap beer and "hacked" fountains overflowing with bubble bath.

Here was a group of people, after all, whose knowledge of computers and networks was so powerful that its dissemination could lead to an FBI bust. And here also was ample evidence of marginalization, evidence even of a roughly formed anti-authoritarian political ideology—scrawled in T-shirt slogans, cobbled together in websites arguing the innocence of jailed hackers, imprinted even in that basic hacker query *What else can this do?*

Before I could fully sketch out the map of the glorious utopia to be built by this elite cadre of politically conscious hackers, though, my train of thought was interrupted by the crackling laugh of the cDc's stuffed-cow mascot. Someone had rigged it so that, when you turned it on and it started to cackle and leap around, it appeared to be sodomizing a stuffed Barney doll. The very same, um, passion play would be acted out again at the end of the Hacktivismo panel the next day, to a big, tension-breaking laugh. It was almost like the cDc was making amends for bringing such a downer topic into the proceedings, giving the crowd a wink that said: This is still DEF CON, after all.

2

POP CULTURE REBOOTED

I remember seeing a story about *The Simpsons* writing team once in which one of the show's writers lamented that the best minds in the Ivy League were all fighting each other to get jobs writing dialogue for cartoons and sitcoms. More recently, I've come across similar sentiments regarding a generation's finest being harnessed primarily in the development of new smartphone apps. Throw in the billions made and lost and made again by the shrewd and well-educated on the trading of obscure financial instruments, and there's a strong case to be made for the first years of the twenty-first century as a singularly frivolous time, a disaster borne of chuckling distractions. If only there were enough of us with the wherewithal to even learn to fiddle while Rome burns; mostly we're just satirizing that old-timey fiddling, posting the results to YouTube, developing a killer fiddle-simulator app and financing it by betting against a brighter future.

As a sporadic reporter on the pop culture beat, I attempted to strike a balance between lamenting this strange state of affairs and celebrating the moments of real creativity and progressive intent amidst the fluff. I never (fully) bought the notion that digital technology was intrinsically less worthwhile than classical culture, and I categorically reject the idea that mainstream mass culture is "lower" and less artful than highfalutin highbrow stuff. I still like *The Simpsons* better than Dickens, and I think Homer had more to say about the cultural chaos of the last twenty years than Oliver Twist or Nicholas Nicholby ever could.

As one of the last children of the pre-digital age entering the workforce just as new technologies rewrote the source code of the whole culture, I had front-row seats for the kind of calamitous upheaval in popular culture last seen perhaps as far back as the days of Gutenberg. And the printing press took centuries to conduct its reboot of western civilization. The internet and its digital brethren did so in just a few wild years.

Like many young writers, I thought at one point I wanted to review music or movies or something like that. I even had a regular gig churning out online film reviews in 1996, which made me a pioneer in a medium I hadn't even begun to understand. In time, though, I decided there was much more to be said about a pop cultural product than whether it was good or bad. I kept notes as best I could, and I tried to tease out connections between the disparate eruptions of a pop revolution. I was also, for a time, one of the first non-technical people in Canada to be paid to play video games. There are countless worse ways to make (part of) a living.

Video games, pop songs, viral videos, soda pop brand names, even tabloid headlines—these are the shorthand of this time, the vernacular of a self-aware, overstimulated, ultraconnected, hyperconsumerist age. This is a real beat, maybe not just despite but because of its ephemeral nature, and one well worth covering. And as with the digital frontier, it's a beat that still benefits enormously from being there, making the pilgrimage and bearing witness. It's not enough to simply know Pepsi once tried to peddle a breakfast drink; you have to go meet the kook who salvaged a can of the stuff from the dustbin of history.

WHY TECHNOLOGY IS FAILING US
(AND HOW WE CAN FIX IT)

Shift, September 2001

It's fair to say I underestimated the transformative power of mobile internet technology. Otherwise, I pretty much stand by my thesis: If fossil-free renewable energy were as high a priority on the global public agenda as the iPhone 5 or Facebook's IPO, we'd already be halfway to solving the existential crisis of climate change by now. Meantime, check out this new Roman fiddle app I just downloaded.

I. A GODDAMN MIRACLE

I WANT TO TELL YOU ABOUT something that happened to me last fall.
I want you to know about it, and I want you to know what it meant.
This is important. You have to understand. It's important because it
was a miracle of modern technology. It was—listen—it was a goddamn miracle.

It happened at the Gilles-Villeneuve Circuit, which is a racetrack
in Montreal. It's on Île Notre-Dame, this racetrack. This is where
the Canadian Grand Prix is held; I've heard that people use it for
rollerblading at other times. It's right out in the middle of the St.
Lawrence River, and on blustery fall days the wind blows across it
and cuts right through you. The island—Île Notre-Dame—is artificial, built for Expo 67 by dumping 25 million tons of earth into
the river. It's the subject of one of those Heritage Minutes—maybe
you've seen the one. I was there for the 17th International Electric
Vehicle Symposium. EVS-17. It was a Monday, mid-morning, over-
cast and unseasonably cold. I'm telling you all this because I want
you to understand. This was not a dream or an idea or a theory. This
exists. It happened. I was there.

Here's what happened. There was a long line of cars and trucks
and smaller vehicles that looked like golf carts, all lined up along a
strip of blacktop that was, I guess, the pit area. The media had been
invited there to test-drive some of EVS-17's finest vehicles. There
was a handful of people buzzing around each vehicle—adding to

the whole pit-crew vibe—and when you got to the front of the line
at each one, an engineer would get in the passenger's side and you'd
hop in the driver's side, and you'd take a spin around the track. Just
like that. Yes, and right up at the front of the line of cars, there was
a white Ford sedan, a car called a Ford Contour. (Here are some
things that *Motor Trend* magazine said about the Ford Contour: "a
strong entry in the family-friendly sedan class" and "offers fun-to-
drive goodness at an affordable price.") The Ford Contour at EVS-
17 had a slightly longer-than-normal body, and it went by the name
of Ford P2000. And when I got to the front of the line, that's when
it happened.

There was a small crowd gathered there, a line of reporters in a
loose semicircle at the rear of the car. They were all looking down at
a man in a navy blue Ford windbreaker who was squatting next to
the Ford P2000's exhaust pipe. Later, this man would sit next to me
as I drove the Ford P2000, bearing witness as I learned that the car
worked just the way you'd expect a strong entry in the family-friendly
sedan class to work, and he would hand me a card that identified
him as Ronald D. Gilland, a propulsion systems engineer at Ford
Research Laboratory in Dearborn, Michigan. At this point, though,
this Ronald Gilland was simply squatting there, making a joke about
the weather. The P2000 was running, even though you couldn't hear
it, and thin vaporous clouds puffed out of the tailpipe next to him.

Someone asked about those clouds.

Gilland said, "It's pretty much like the outflow on a clothes dryer."
He said it just like that, nonchalant and conversational.

And then he held his hand in front of the exhaust pipe for a few
seconds. And then held it up for inspection. We leaned in, all of us.

His hand was dotted with drops of water.

Do you need to know that the P2000 is the prototype of a car
that Ford plans to make one day? That its motor is electric, and that
that motor gets its power from a hydrogen fuel cell? (Specifically:
the Ballard Mark 700 Series, a proton-exchange-membrane fuel
cell made by Ballard Power Systems.) Is it important that there is
a hydrogen tank in the P2000's trunk, and an air compressor up
front, and that when the hydrogen and oxygen combine in the fuel
cell, they generate electricity? Do you need to know the P2000's

wheelbase or curb weight or peak torque? No. I don't think you do. Not really. The important thing—the goddamn miracle—is that when this guy put his hand in front of the exhaust of this fully functional automobile, the only thing that it wound up covered in was drops of water.

When I want to feel good about the future, when I want to believe that we are moving toward a better world, this is what I think about: drops of pure water on a Ford engineer's hand, water from the exhaust pipe of a Ford sedan. Water clean enough to drink.

II. YOU CAN STUFF THAT WEB-ENABLED CELLPHONE RIGHT UP YOUR VACUOUS ASS [I]: THE FUTURE IS FRIVOLOUS

DO YOU REMEMBER THOSE AT&T ads? The ones that ended with the tagline *You will?* You know: They were shot in this near-future utopia that was supposed to look all sleek and clean but came off—at least from where I was sitting—Blade Runner-esque and kind of creepy. They ran from the spring of 1993 to late 1994. TV, radio, newspapers and magazines. They won a bunch of awards. They'd start off with a voice-over, the intonation of a rhetorical question. *Have you ever...?* And then there'd be these scenarios you never dreamed of—someone's entire grocery cart checked out all at once, a door unlocking to the sound of someone's voice, someone sending a fax from a beach. I found an old business-news article about these things that called them "soon-to-be-released technologies." (I also discovered that it was Tom Selleck who did the voice-overs. I hadn't remembered that.) And there was one, remember, a TV spot that showed someone in a phone booth, and the voice-over said: "Have you ever tucked your baby in from a payphone?" And then a pause, and then a shot of a baby sleeping contentedly somewhere else. The voice-over: "You will." Chilling. Almost Orwellian. You will.

Funny thing: Turns out those ads—the ones on TV—were directed by David Fincher, who went on to direct *Fight Club*, a movie that was about destroying the companies responsible for creating clean, sleek facades, among other things. Suddenly those AT&T ads make a little more sense. The company thought they'd make everyone all weak-in-the-knees for gizmos and gimmickry, I'm sure. But maybe Fincher wanted us to be a little bit afraid.

We weren't afraid. Not even a bit. Or if we were, we talked ourselves out of it. So we never really considered whether or not we wanted any of the stuff we were making. Whether or not, for example, it solved any discernable problem in our lives. Or in our world.

All we've done, it seems, since around the time of those AT&T ads, is talk about the future. The future, and how different it would be, and how amazing, and—it was implicit—how much better. And we got so wound up in all this random speculation that we never did get around to figuring out what we needed; we just looked for the stuff with the greatest gee-whiz value—Tamagotchis, email pagers, customized web portals that remember our names and hobbies—and that's what we built. The latest thing—and you've probably heard all about this—is using your cellphone to surf the web. The "mobile internet." It's all the rage in Japan even now. I can't even begin to estimate what it'll cost to make it possible for each and every one of us to be able, if we so choose, to use our cellphones to browse the web. But to give it a stat: Openwave Systems, the self-proclaimed "world's largest provider of mobile internet software," had about US$556.1 million in total operating expenses for the six months ended December 31, 2000. And I'm quite sure it just builds from there.

I've got nothing against Openwave per se, and $556.1 million has certainly been spent in more pointless ways. (The many failed prototypes of what is currently referred to by American leaders as the "National Missile Defense" system come to mind.) The point, though, is this: It was impossible even to conceive of using a cellular phone to look at web pages in 1990, the year the Intergovernmental Panel on Climate Change (IPCC) issued its first report on global warming. Even in 1995, the year that the IPCC asserted that global warming was a reality, there was essentially no demand for web-enabled cellphones. Why, then, is there such a rush—and such vast reserves of cash, talent, time and wherewithal—to bring web-enabled cellphones to market? Why, that is, have we put such an enormous amount of our resources into developing new technologies to solve problems that were only called into being by slightly older technologies?

Put another way: There is a trajectory, an arc that you can trace from the AT&T "You Will" ads to Openwave's half-a-billion dollars in expenses. And it misses completely the greatest problems of our time. One in particular.

III. THE PROBLEMS THAT I ALLUDED TO JUST NOW

HERE'S WHAT'S BEING MISSED: a cluster of problems that I'll place under the rough heading "environmental degradation." I'd hate to imply that any one aspect of the process by which we are making our planet unfit for human life is more troubling than any other, but the one in particular—the one that should really be keeping our engineers and genius inventors up at night, working on solutions—is global warming.

Can I take it for granted that I don't need to tell you why the degradation of the environment is the biggest problem of our age? That it is the threat to our livelihood—the World War, the Great Depression, the would-be Nuclear Winter—against which we need to mobilize the full power of our resources? I would like to think I can take this for granted.

I'd like simply to assume that you know that it—this degradation, this destruction, this systemic poisoning—supersedes the current or near-future state of any national economy. That it is an unfolding calamity far greater than a wave of new tensions in Sino-American relations or another round of violence in the Middle East. That it is not an "issue" the way, say, the balance of powers between federal and provincial governments is an "issue." That it is a cluster of events—events resulting from human activity on this planet—that are demonstrably, measurably happening. That it is not, therefore, an ideological construct. That while it might be possible to assemble an argument or voice an opinion about clean air and water, and fertile soil, and a habitable climate, that these opinions are not right or wrong so much as utterly irrelevant. That, for example, the sun's ultraviolet light, when it reaches the earth without being filtered through a layer of ozone, is capable of producing malignant melanoma in the skin tissue of any person, totally regardless of that person's opinion about the relative importance of "environmental issues." Can I take all of this for granted?

Maybe I can't. These are, after all, details that get reported with far less regularity than the fluctuations in Nortel's stock price. Should I assemble some of the evidence? The polar ice cap is melting, as are the snows of Kilimanjaro. Last November, the leaders of thirty-nine of the world's small island nations petitioned the UN to take action on global warming before rising oceans swallow up large parts of their countries. Is that enough? Last September and October, officials in Punta Arenas, Chile, advised the city's 120,000 citizens to stay indoors from 11 A.M. to 3 P.M. to avoid the sun. At present, approximately 19,000 Ontario lakes have acidity levels inimical to plant and animal life due to abnormally high sulphuric acid levels in the rain that falls into them. The hole in the ozone layer over Antarctica is just under three times the size of Canada. The government of Tuvalu—one of those island nations worried it might be submerged by the rising oceans—has looked into buying land in another country, just in case.

All this is really happening. It's bigger than you. It needs to be fixed. And there are ways to fix it. That's my point.

IV. MORE MIRACLES OF MODERN SCIENCE THAT YOU MAY NOT HAVE HEARD ABOUT

BEFORE THE END OF THIS YEAR, the majority of the Stateline Wind Generating Project on the Washington-Oregon border—the world's largest single wind-energy development, 450 turbines in all—will begin creating power for 70,000 homes. The world's first commercial wave-power station went online on the Scottish island of Islay last fall. A project is underway in southeastern Spain to build the world's largest photovoltaic power plant—four times larger than any other solar-power facility currently in operation. A gas station in San Francisco began selling biodiesel—a fuel made from food oils that can be used to run any diesel engine—in May. These are just the renewable-energy solutions, and they were not hard to find. Dig a little deeper, and you'll find not just constellations but whole vast galaxies of green technologies. Go, for example, to the Environmental News

Network website (enn.com) and click on "Marketplace." The site's links number in the hundreds: cottage-industry consumer products, business services, green versions of every single item at your average supermarket.

Pick one at random, though—say, SolarRoofs.com, distributors of the Fireball 2001 solar water heater—and it may become clear why not all of these ideas have made it to your local mall. The idea is sound enough: install a solar panel on your roof to heat your water. Pays for itself in just a few years. Sign me up. Except look at the website. Look at the Fireball 2001's cheap, messy website. It's horribly designed, cluttered, littered with grammatical errors and misused quotation marks. It comes across about as legit as an infomercial for some miracle rust remover. Not very many homeowners are going to make enormous changes to the way they get their hot water based on a sales pitch that reminds them of an infomercial.

In fact, isn't all this stuff—these half-formed and poorly executed bright ideas, these small businesses without the money or the business acumen to make it out of the planning stages, these ingenious inventions and revolutionary engineering feats waiting to be market-capped—isn't it a little like, say, if you'd gotten lost driving into San Francisco twenty, thirty years ago, and found yourself out in the middle of the farmer's fields of the Santa Clara Valley? And—who knew?—there's all these R&D facilities full of ambitious geeks with big ideas out there. Isn't it, in a way, kind of like Cisco and Intel and Yahoo and all the rest are just sitting there, waiting to be scooped up and driven hard into the mainstream?

V. WHY THIS MIGHT NOT BE ALL THAT FAR-FETCHED

THERE IS A HIGH-PROFILE BUT somewhat superficial reason to posit the idea that green tech (for desperate want of a better catch-all term) could become the elusive Next Big Thing in the high-tech world. That reason is this: Both Bill Gates and Paul Allen have invested heavily in renewable energy companies. Also, like the various communications technologies before them, green technologies have the potential to create an enormous re-ordering of the business world. "I believe fuel cell vehicles will finally end the 100-year reign of the internal combustion engine"—that's how one starry-eyed evangelist phrased it.

Here's something, though. Here's the real reason this hypothesis isn't so farfetched: that starry-eyed evangelist was William Clay Ford, Jr., chairman of the board of Ford Motor Company and great grandson of Henry Ford. He said that to a gathering of the automotive world's elite in January 2000. He told them some other unconventional things, too. He said that consumers are smart enough to know that big corporations have the money and power to deliver the long-promised "better world." He said that sustainability and social responsibility would soon be essential to success in the corporate world. He said, "This is a trend that is not going away." It was a remarkable speech. And then, this past February, DuPont—I mean, *DuPont*, for god's sake—formed a fuel cell business unit. The press release reads: "DuPont is seeking a strong presence in what it believes will be a $10 billion total market for fuel cells by the year 2010."

The Economist—another gang of wild-eyed ecoterrorist radicals, to be sure—speculated in April that "energy technology" might just be "the next big thing" (their phrases), that it was the "one corner of the technology world" that might escape the bust.

Doesn't it sound inevitable, this change? Kind of like the birth of a new medium?

VI. WHY IT MIGHT BE FAR-FETCHED AFTER ALL

A SLIGHTLY LONGER VERSION of that quote from Ford: "*Longer term*, I believe fuel cell vehicles will finally end the 100-year reign of the internal combustion engine" (emphasis mine). Of course.

After all, that fall day in Montreal when I drove the Ford P2000 was more than six months after Ford's speech, and still the guys in Ford windbreakers were all talking five, seven, maybe even ten years before this thing's ready for the mainstream. Oh, they might have to produce a few by 2003 to meet California's new zero-emissions guidelines— they told me things like this with a kind of winking skepticism, like they figured California wasn't *really* going to enforce those new guidelines that stringently—yeah, they'd have to make a couple, but this was mostly a future-tense thing. (The California government's legislation requires ten percent of all new cars sold in the state from 2003 forward to be zero-emissions vehicles.) Right now, though—well, be reasonable. Fuel cells are too expensive to make, it's all too expensive, and hydrogen explodes easily, and—above all the rest—where are you going to get enough hydrogen for hundreds of thousands of cars, anyway? Later, I talked to Firoz Rasul, Ballard's chairman and CEO, on the phone, and he told me I'd see fuel cells in homes—in generators, that sort of thing—before I'd own a car that ran on one.

It's funny about all this: I could show you press clippings from 1992—before DaimlerChrysler and Ford and Toyota and Honda all invested millions in fuel cells—that make the exact same arguments.

Another funny thing: At EVS-17, I walked across the showroom floor to a booth you could literally see from the big displays where Ford and the rest were displaying their future-tense cars. Here was Stuart Energy, a little company that has been making hydrogen generators for years. In fact, when Ballard tested its fuel cell–powered buses in California, Vancouver and Chicago, it was Stuart that set up the hydrogen filling stations. The Stuart representative showed me a sleek box about the size of a dishwasher: the Stuart personal fuel appliance. The way the Stuart rep told it, if there were, say, 300,000 fuel cell–powered cars being sold in a year, Stuart could amp up its production and bring the cost down to the point where one of these Stuart boxes—which plug tight into the same kind of outlet as a clothes washer—could be sold affordably along with the cars. Plus, Stuart also makes neighbourhood-sized generators, so someone could set up some kind of new-fangled store in a central location and then customers could come by as needed to fill up their cars with hydrogen, perhaps using some sort of pump-like apparatus. A revolutionary concept, to be sure.

I mean, *fuck*—is it really all that easy to get gasoline from the ground to your SUV's gargantuan tank?

Lesson learned: Car company executives may talk revolution, but another reason they're so heavily invested in companies like Ballard is so they can control the rate of change. And they'd like it to happen nice and slow, thanks.

Actually, it's kind of foolish to think that the likes of Ford, DaimlerChrysler and DuPont—not to mention BP, the corporation formerly known as British *Petroleum*, one of the world's top producers of solar power—to think that these hoary old oil-dependent companies were going to lead a revolution anyway. After all, it wasn't AT&T and Bell Canada that set the pace in the telecommunications revolution, now was it?

VII. THE $135-MILLION BOO.COM SOLUTION

So here's the situation: We have some spectacular new technologies that a wide cross-section of humanity considers worthwhile and beneficial and would thus likely buy, except that they are currently too expensive and niche-marketed and generally fringe-oriented to be viable as businesses. This, as I understand it, was roughly the case with the internet circa 1992. (Well, there is one exception: There was no demonstrable mainstream demand for internet technologies circa 1992.) And then, of course, the whole thing exploded, and nonsense like Boo.com happened.

I have no doubt you remember Boo.com, the British online retailer of hipster threads that became the dotcom bubble's most spectacular bust and the most widely used metaphor for marketing folly since the Edsel. It was a helluva ride: Nearly US$135 million in venture capital burned on a website that stayed live a mere seven months; more than 400 employees and US$42 million spent on advertising in order to sell almost nothing; US$200,000 per month spent just on travel expenses.

Easy cautionary tales aside, I think the really critical quality that Boo.com embodies is the infectious power of human exuberance. Boo.com makes it abundantly clear—in such sensational, panoramic detail that there should never again be even the tiniest shred of doubt about the matter—that there is absolutely nothing inherently rational or reasonable about our economic system. That, quite the contrary, it frequently gets taken for wild but nevertheless productive rides by decidedly irrational forces like hype, excitement and sheer desire.

Of course, I am not an officially sanctioned business leader or economist, and one of the favourite tricks of such people, when they stumble upon ideas that run contrary to their beloved status quo, is to dismiss their viewpoints as ignorant of the infallible science of economics. So I'll turn once again to voices from the corporate ranks: Brian Milner and Patrick Brethour, both business writers for the *Globe and Mail*. In a joint op-ed piece in March, the pair wrote a lengthy thank-you note to the greedheads of the dotcom bubble, arguing that said bubble was not folly at all but rather an essential foundational step in the "twenty-first-century communications revolution." Why? Because, as they put it, "you cannot cross a chasm in two leaps." The internet needed a great big initial jump, and the speculative fever of the past couple of years—the fever that allowed companies like Boo.com to scare up $135 million in a few short months on little more than a vague idea and some charm—made that jump possible. Some folks may have taken a bruise or two on the landing, but now the crucial infrastructure is in place. And note in particular that Milner and Brethour mean more than fibre-optic cable when they say "infrastructure." The term, they say, also includes "the intellectual legacy of the dotcoms, whose explosive push into online retailing, wholesaling, information and business services has forced the occupants of the bricks-and-mortar world to go along for the ride."

This is the spirit that the green-tech companies need to tap into. Most of the dotcoms were nobody's idea of a sound business model, but taken as a whole they represent a fantastic *social* model. Just imagine that reckless energy pointed in the direction of a real problem.

One more of those funny things: On the very next page of the issue of the *Globe* in which Milner and Brethour's essay appeared, there was a lead editorial about the American government's decision to back out of the Kyoto protocol.

VIII. YOU CAN STUFF THAT WEB-ENABLED CELLPHONE RIGHT UP YOUR VACUOUS ASS [II]: THE SYMBOLS OF SILICON VALLEY

DOESN'T IT SEEM, AFTER ALL, like the current technological "revolution" could use some direction? I mean, does it actually stand for anything at this point? Is it a revolution at all? But—fine—let's take the rhetoric at face value. Let's assume that something revolutionary has happened. And so let's go to the revolution's heart—to Silicon Valley. Let's go there and look for symbols, and ask what Silicon Valley *means*.

What are the symbols of Silicon Valley? Do you know? Forget abstractions and commodities. Lines of binary code and microchips— these are the things you see used to illustrate stories about Silicon Valley, but these can be made and used anywhere. If we are talking about revolution, then we are talking about something more fundamental: about competing ways of life. And so we must also ignore the haunts of the very few at the top of the Valley's food chain—Buck's Restaurant, for example, the legendary breakfast joint where venture capitalists take meetings to decide how to disperse their millions. There's simply not enough room at Buck's for everyone to sit after the revolution has been won. It can't encapsulate it all.

So instead consider Highway 101, Silicon Valley's primary north-south artery, the great strip of blacktop that ties its constituent parts together. At rush hour, the bumper-to-bumper traffic on Highway 101 often stretches for miles. There are your revolutionaries, one per upscale sedan and SUV, the carpool lanes (for cars with two or more passengers) left mostly empty. Suburban sprawl stretches away to the walls of the Valley, covering over soil rumoured to be as fertile as the Nile Valley's. And the revolutionaries sit in their cars, barely moving, under a clear blue California sky turned yellow-brown by the exhaust from their cars.

And consider also San Jose. It proclaims itself "The Capital of Silicon Valley" on banners at the airport; it is the closest thing the Valley has to a city. At San Jose's northeastern edge, where it bleeds imperceptibly into Milpitas, there is a long, curving boulevard called Tasman Drive, and all along either side of it are low-slung office buildings—most of them two or three storeys tall, none more than six. Office parks, one after another, each building surrounded by a wide apron of parking lot, Cisco is the biggest presence here— literally dozens of Cisco buildings, all of them uniform pale-beige. From the heart of the Cisco ghetto, it's about a ten-minute drive at 80 km/h—over on Tasman and then south on North First Street— until you hit something that isn't an office park. At the southwest corner of North First and Trimble Road, finally, you'll find a small strip mall. There is a Carl's Jr. fast-food restaurant here, Mexican and Indian restaurants, a coffee shop, an optometrist's office. There is a mass-transit train running down the centre of North First, but almost everyone drives here. And so at lunch, there's quite a scene at this strip mall: the parking lot packed, cars and an inordinate number of hulking SUVs circling the parking lot—circling slow, lumbering, too big for the tight turns of the strip-mall lot—all these vehicles jockeying for parking spots. These are the same people who'll be out on Highway 101 later the same day in those same vehicles under the smog-browned sky, and this is how they live: behind the wheel of their SUVs, fighting for parking spots at a strip mall. They look like cattle pushing and shoving each other as they pass through a maze of fences and gateways, and how could they know about the slaughterhouse that awaits?

This doesn't look like much of a revolution.

IX. THE REVOLUTION WILL NOT BE ONLINE

THE TREMENDOUS AMOUNTS of time and energy and resources we have devoted of late to working out the kinks in a new mass-communications medium—whether or not it amounts to a revolution, this process has provided us with an excellent blueprint for how to instigate and manage enormous, rapid change. It has even mobilized some of the right kinds of skilled labour and capital to execute it. (The kinds, for example, that are adept at dealing in innovative solutions and aren't beholden to enormous oil and gas companies.) And the really great thing is, much of the groundwork has been done. We have invented solutions but haven't implemented them. That's why I keep coming back to the Ford P2000 and the water steaming out its tailpipe: because the damn thing is just sitting there, a finished product, waiting.

See, because maybe if those lines of cars out on 101 in the Valley had all been dripping water our their tailpipes, and maybe if the carpool lanes and electric commuter trains had been full, and maybe if there were solar panels on the roofs of those office parks and wind turbines in the vacant lots—well, hell, maybe then it would've looked revolutionary.

And because, most of all, these are the things we *actually* can't live without. Peer-to-peer technology, the wireless web, Super Bowl commercials starring sock puppets—the relative merits of all of these are open to discussion. Here's something that isn't: the absolute, bottom-line necessity of clean air, potable water, fertile soil, climactic conditions favourable to human survival. It's not debatable, not something to be put off till we all have more time, not a luxury or a lifestyle choice. Surely you understand that. This is the revolution we need.

THE SGT. PEPPER OF GAMING

Shift, April 1999

This is likely one of those things that seems inevitable in retrospect—that video games would, in some significant measure, come to occupy a space in the cultural hierarchy of similar import to that of a movie or pop record. Just a little over a decade ago, though, much of the mainstream media still regarded video games as irrelevant children's playthings. There was something almost thrilling about exploring them in depth in issue after issue of Shift—*a psychic journey into uncharted terrain, where not even the basic criteria for review had been established.*

I'm sure I got many details wrong documenting this virgin territory, but I did make two early calls that have stood the test of time. First, after attending the 2001 E3 show in Los Angeles (the video game industry's premier trade show), I knew Rockstar Games and their Grand Theft Auto *franchise would take the world by storm. And second, as catalogued in the column below, I was certain* The Legend of Zelda *would become the aesthetic benchmark for all that followed.*

His name is Shigeru Miyamoto, and he is, by the few reports extant, a humble man, unassuming, modest, perhaps even a touch self-deprecating. His background is in industrial design, and his career at Nintendo began with the mundane task of designing cabinets for the company's arcade games. In 1981, he was given a shot at mapping out an actual game—*Donkey Kong*—because, he says, "No one else was available."

Well, lucky Nintendo. The game was a major hit, first at the arcades, then for ColecoVision and the first-generation Nintendo console. So Miyamoto designed another game, a sort of sequel and expansion of *Donkey Kong*, called *Mario Bros.* That was a hit, too. Then he designed other games—several spinoffs of *DK* and *Mario*, a couple of flying games and an adventure called *The Legend of Zelda*—and, whaddaya know, they were all hits, too. Big ones.

Fast-forward to 1998, and this Miyamoto-san is now general manager of the Entertainment Analysis and Development Department at Nintendo, a dry, corporate title which masks the fact that the two most successful video game franchises ever created (the *Mario* and *Zelda* series) were born primarily of his imagination. Miyamoto likes the mask. Though Nintendo has tried to trump him up as its "star" designer and hardcore gamers have started to treat him as something of an idol, though *Next Generation* magazine has gone so far as to dub him "the Spielberg of video games," Miyamoto demurs. He only wants "to be known in the industry," nothing more.

So, yes, his name is Shigeru Miyamoto, and I guess he won't like me for saying it this way, but he has invaded my dreams.

Last night, for example, I dreamed that I was swimming in the clear, cool waters of Lake Hylia, diving off cliffs and under cascading waterfalls. It was the most beautiful swimming hole I'd ever seen, and the most refreshing dip I'd ever taken. It's also a geographic region not of this earth but of Hyrule, the mythic world brought to impossibly vivid life in *The Legend of Zelda: Ocarina of Time*, Miyamoto's latest and greatest imagining of the *Zelda* world, the first for the N64 and fifth in the series, and just a staggering achievement of game design and engineering. *Zelda 64*, as its legions of fans usually refer to it, is a quantum leap forward in the medium: It single-handedly catapults the whole video game industry from the kid's-stuff world of entertainment into the more sacred, sombre halls of high art. It is, in short, the *Sgt. Pepper* or *Citizen Kane* of video games, the work that finally, triumphantly, combines all of its medium's best features—the breathtaking graphic detail of *Myst*, the cinematic lyricism of *Resident Evil*, the brain-melting logic puzzles of the best adventure games (including previous *Zelda* games), and sheer giddy playability of, well, the *Mario* series—into a single artistic statement. Years from now, when the notion of video games as art and game designers as auteurs has been firmly established, Shigeru Miyamoto and his *Ocarina of Time* will loom Welles-like over the industry.

Zelda 64 launched last November to ecstatic critical praise (I haven't seen a single review that *doesn't* call it the greatest video game of all time) and unprecedented sales. With a ticket price more than triple that your average music CD, *Zelda 64* sold more than a million copies in its first two weeks release and 2.5 million in the first six. What's more, it has become a sort of "killer app" for the N64; that is, people are dropping bucks on the platform just to play it. Indeed, Nintendo 64 platform sales shot up forty percent in the four weeks following the game's release.

All of which begs the question: What's so damn great about it, anyway? Well, the sheer lushness of your environment—of Hyrule— is what strikes you first. Every leaf on every tree is fully rendered. Your character, a forest-dwelling elf named Link, moves with preternatural fluidity, and the fingers on his hands are clearly visible when he climbs a fence. Hyrule has sunrises and sunsets; some mornings, the field outside the castle is shrouded in ground fog. And when you swim Zora's River—even at night—your wake dances around you in widening concentric rings. Then, just as you get used the beauty of the place, your jaw drops yet again when you catch another layer of detail—say, when you use your full 360 degrees of vision to watch a Hyrulian sun paint the sky in oranges and pinks, and light spots appear on the screen when you pan past the sun itself. Later, once you've oriented yourself, it might be the complexity of the tasks and the clever twists of the plot that have you gaping. But finally, maybe after a day or two, you won't be amazed by what you're looking at, because you won't be looking at it at all. You'll be *in* it.

It's this last quality that begins to answer the bigger question about what kind of aesthetic the best video games prescribe. Because, for all its visual pyrotechnics, *Zelda 64*'s (and, at this early stage, the medium's) most distinctive artistic achievement is its immersiveness. From film, *Zelda 64* has borrowed perspective. Most of the game is "shot" like a movie, with Link moving into a frame at one point and exiting at another, then trotting into a new frame that might be tighter or shot from above instead of ground level. The game's plot owes a large debt to the novel, in that the events and experiences at a particular stage in Link's "story" only make sense in the context of previous stages (though it warrants mention that if *Zelda 64* has a glaring weakness, it is its plot, which relies—like the vast majority of its video game predecessors—on stock narratives and characters).

But its immersive quality is where the game—and the medium in general—stands on its own aesthetically. This is in part a function of time. Even the best movies are sometimes labelled "too long" if they run more than three hours. Ditto for albums half that length. And while some novels may demand a considerable time commitment, it is usually in small parcels. *Zelda 64* requires an investment of sixty hours at the very least, yet it's invariably harder to pull yourself away from the game than to keep playing, even after a twelve-hour session.

Still, there's more to it than simply the number of hours involved. The game—and every player's experience of it—is made unique by the fact that it encourages romping. There are certain guidelines to your quest, yes, certain places you need to go next, but these don't stop you from spending a few hours fishing at Lake Hylia or shooting targets (which, incidentally, exist in the form of red, green and blue crystals that explode in a rain of twinkling shards, a stunning example both of *Zelda 64*'s lyricism and its depth of detail) at a gallery in the market. Indeed, such detours are often required—catching a record-breaking fish will earn you a prize that's essential to making it through the next stage of the game.

Zelda 64, for all its merits, is not flawless. The soundtrack, for example, quickly grows cloying, and the successful cross-breeding experiments in my living room with ambient techno, trip-hop, dub and, most impressively, vintage Floyd suggest that someone needs to take up the task of properly scoring worthy games. The game earns adulation, though, as much for the potential it has unlocked as for what it has itself accomplished. Like *Sgt. Pepper*, it reveals the boundless possibilities of its medium. Miyamoto has got me dreaming inside his game's milieu. In fact, *Zelda 64* made me want to run out and start designing games myself, and I hope and trust that other game designers will note this raising of the bar and push the medium further.

Now if you'll excuse me, it's getting late, and I have some romping to do. I still haven't conquered the Ice Temple, and that missing key has got to be tucked away in there somewhere.

GAMES WITHOUT FRONTIERS

Shift, October 2000

I'm not sure what it says about me—or about India—that the closest I came to reporting on the place directly was to describe it as an analogy for being stuck inside a video game. Maybe that both can seem to an outside observer to be characterized by a kind of sensory overload that makes them difficult to report on in any way that feels both resonant and comprehensive. I could tell you exactly how my game went, but why would you care? Or I could try to extrapolate and generalize, and get so much of it wrong.

BASGO GOMPA SITS ON A BARREN, rocky promontory overlooking the Indus Valley in Ladakh, high in the Indian Himalaya. It is a sprawling complex of sacred temples and monks' residences, and it has stood stoic watch over the valley since the early 1500s. Even today, though half-collapsed, the monastery is an awesome sight set against a towering wall of jagged Himalayan peaks, especially when viewed from the elevated vantage point of the roadside west of the village. We approached Basgo from the east, though, where the views aren't so panoramic, and so we had nearly bounced past it in our rented jeep before one of my companions noticed it and asked the driver to stop. We followed a series of narrow laneways up through the village, got further directions from a grinning Ladakhi woman with a gaggle of giggling kids at her feet, and pushed on. We dodged a herd of sheep, following zigzagging footpaths that wound around to the other side of the promontory, and passed abandoned outbuildings and ruins. We encountered locked doors and dead-end passages, backtracked a couple of times, and had to traverse a narrow ledge at one point, the valley floor hundreds of feet below. Finally, we came to a sort of large balcony. The monastery's main building towered several stories above us, but we couldn't find a way in.

I noticed a broken window, its pane opaque with dirt, and wandered over to see if it would reveal any clues. The hole in the window was at about chest level, so I stooped a bit, peered in...and found myself staring into the eyes of an enormous Buddha. The statue's body disappeared into darkness two storeys below, but the bright golds and blues of its head were stage lit by shadowy sunlight. After I got over my initial surprise—I had stopped breathing for a good ten seconds—I started to grin, thinking: *And, of course, now the fire-breathing dragon comes. Defeat it, and the Buddha gives us the keys so we can get to the next level.*

I mean, it just made sense: The quaint helpful villagers, the dead ends, the tests of dexterity and the great prize now just inches away. Never mind that we were at least three days away from the nearest arcade, or that it'd been several months since any of us had even seen a Nintendo console. As far as I was concerned, the whole experience resembled nothing so much as a video game. In fact, a key sequence in *Tomb Raider II* is set in a Tibetan monastery, with the critical piece of information tucked behind the head of an enormous Buddha.

But this was, of course, a *real* Buddha, the centrepiece of a flesh-and-blood-and-spectacular-vistas experience, so I felt a little silly, at first, about equating the great Buddha with anything even remotely connected to Lara Croft. No, not just silly—I felt that there was something... wrong with the intrusion of video games into this decidedly un-digitial experience. Surely, this is a cautionary tale about the myriad ways in which our hypermediated society invades our consciousness, impeding us from the undiluted enjoyment of even our purest and most sublime experiences. I mean, comparing the serene eyes of a colossal Buddha with the grotesque little pixels of a video game? It's like going to Stonehenge and being able to think only of that scene in *This Is Spinal Tap*. It's an outrage! Alert the authorities!

The problem wasn't that I wanted my experience at Basgo Gompa to be entirely unmediated (a concept at least as fanciful as fire-breathing dragons); it was that I was still infected by the notion that video games were too trivial to inform the event. We are strongly encouraged to see video games as nothing more than childish diversions, far too lacking in substance to contribute to our understanding of the world the way that a Tennyson poem or a Scorsese film might. Like comic books and cartoons before them, video games are relegated to the ghetto of the inherently juvenile. And something inherently juvenile does not inform your aesthetic; it merely degrades it.

So what to do, then, with the realization that video games are, in fact, informing the aesthetics of our culture at large? It's not just me on a hillside in the Himalaya; the form and content of video games have begun to spill over into all manner of other media. Consider, for example, Alex Garland's novel, *The Beach,* in which video games are employed as a central metaphor. To understand how the protagonist, Richard, deals with extreme crisis, look to his reaction the moment before he loses at *Street Fighter.* To understand his approach to power politics, consider his strategy for defeating "bosses" in adventure games. (The film version unfortunately mutates this device into just the kind of hokey cartoon that enables video games to be so routinely dismissed as infantile.) David Cronenberg's 1999 film, *eXistenZ,* went even further; nearly the entire movie is structured as a video game, complete with stilted dialogue, generic settings and plot tics requiring, for example, that the protagonists perform specific tasks or intone certain lines in order to proceed to the next level.

Probably the strongest example, though—and a crystal-clear indication that the video game aesthetic can significantly enhance the meaning and impact of other media—is the video for the Red Hot Chili Peppers' recent single, "Californication." Directed by Jonathan Dayton and Valerie Faris (who collaborated with Todd McFarlane on the award-winning follow-the-bullet video for Korn's "Freak on a Leash"), the video is set in a glossy, soulless L.A. The various Peppers negotiate the bleak landscape in hyper-fast-forward as it crumbles around them, threatening to swallow them whole. The band—and the decaying city—is rendered entirely in the unconvincing 3-D simulacra of an action-adventure game, at least until the Peppers "win" the game by uniting beneath the streets of the collapsed city and shedding their digital skin in a burst of liberating light.

The video is remarkable for its use of the consequence-free joyride of your average action game to illustrate the utterly vacant California of the song. ("Earthquakes are to a girl's guitar / They're just another good vibration," sings Anthony Kiedis, as we watch drummer Chad Smith score a "Bonus" by leaping through the hole in a giant billboard doughnut.) But Dayton and Faris mine this milieu even deeper, using the video game's supposed limitations—blocky graphics, backdrops that scroll by in underdetailed uniformity, the mutation of any and all settings into jungle gyms—to score rhetorical points that far exceed those made by Kiedis's lyrics. The "Californication" clip makes the argument that video game-style graphics are, in fact, a reflection of the bland corporate facade of contemporary America. The tragedy that the video suggests is not merely that we pay too much attention to Hollywood fantasies—it's that we're living inside the stage sets, residing in suburbs as generic as the looped backdrop of a video game, and transforming vacation destinations, shopping malls, Cineplexes and the local bar and grill into theme parks.

Come to think of it, the fact that a Ladakhi monastery reminded me of video games isn't all that troubling. Now, if it had struck me as a really cool environment in which to consume Buffalo wings, *that* would be disturbing.

THE LEGEND OF PEPSI A.M.

This Magazine, November/December 2002

By coincidence, the office of This Magazine *was right around the corner from* Shift's *Balfour Building digs. This felt like* Shift's *more radical and politically astute cousin in those days (as did* Adbusters, *another magazine I wrote for on occasion). In any case, not many magazines will give you a word count this high with which to explore the semiotics of a can of pop (though I had to write a dry piece for a business magazine to cover the travel costs). There were few assignments I took as seriously. The culture of marketing has moved so far and so fast and in so many discombobulating directions since then that the idea of morning cola now seems almost quaint.*

I CANNOT REMEMBER WHEN I first heard of the myth of Pepsi A.M., any more than I can remember when I first drank a Coke. Such is, I suppose, the nature of folklore: it seems to materialize out of the ether, to enter one's consciousness purely by osmosis. I only knew that I knew it, and that it had about it the trappings of a legend.

The tale was invariably told in the broad, get-a-load-of-this strokes of a cocktail party joke, and in tones that suggested it was a paramount example of the folly born of absolute power. It seemed like some post-industrial take on a tale of a feudal monarch's impotence or a federal law-enforcement official's fondness for cross-dressing. The details could not be verified. The evidence had been suppressed. But it gave the peasants a hearty chuckle and spoke to larger truths about the absurdity of our corporate overlords. I mean, Pepsi *A.M.*?

The story goes that a long time ago, possibly even in the 1980s, the Pepsi-Cola Company set about developing a breakfast drink. A carbonated cola, that is, to be drunk exclusively at breakfast—in lieu of, say, coffee. It was lower in carbonation and higher in caffeine than the regular Pepsi, it was alleged. It had been test-launched in select communities in the United States, had failed miserably, and had thus been scrapped, all records of its existence erased.

Its name was Pepsi A.M.

This is what I had heard. And I assumed that it would be all that I would ever know.

But then one fine spring day, I happened upon a news item about an institution that housed the archaeological evidence of failed consumerism. Some sort of museum, I gathered. It had been dubbed, in this story, "the Smithsonian of Stinkers." Its real name was the New Product Showcase & Learning Center (the NPS&LC).

It was not actually a museum, as I learned through a few phone calls, but rather a research and consulting firm catering to enormous consumer enterprises such as Kellogg's, Nabisco, Unilever, even the mighty Procter & Gamble. It was located—quite inexplicably, as if in a Vonnegut novel—in Ithaca, New York. It struck me instantly as the one place on earth that might have hard evidence to verify the Pepsi A.M. myth.

But the NPS&LC was not a public institution; I could not simply arrive at its doors and expect to roam around ogling its treasures as if it were some second-rate tourist attraction, like the Louvre. Its admission standards were far higher than that, and I was decidedly not a Fortune 500 company. Fortunately, I had recently taken to impersonating a business journalist. Before long, I had credentials from one of our national business magazines (the ineptitude of others is an easy sell in any genre), an expense account and a passport into the inner sanctum.

I wanted to see a can of Pepsi A.M. for myself, to hold it in my hands and feel its heft, to read its logos and iconography as a scholar would the pictograms on a cave wall. I wanted to bring the Pepsi A.M. myth out of the darkness of superstition and into the light of truth.

This was my quest.

Cola is among the small handful of commodities used as shorthand symbols for the imperial campaigns of the consumer ethos. It is a top-ranking field marshal in the consumerist army and the soft drink industry is a stellar specimen of the trajectory of the modern corporation. Soft drinks were originally a cottage industry in the true nineteenth century sense. By the mid 1800s there were countless thousands of carbonated tonics and curatives, cooked up in crazed inventors' toolsheds or the backrooms of general stores, some of them on sale in only one small region or even at a single drugstore's fountain. Many relied on opium derivatives for their kick. Then, along came one of the many entrepreneurs of the industrial age. A novelty product was reworked and brought to a mass market. In time, a handful of competitors emerged—other brands, other flavours—but never more than you could count on your hands. Then, at a moment not too long ago, an explosion, like a Roman candle hitting the top of its arc and then scattering a thousand tiny sparks across the empty sky. Out of that fireworks display emerged absurd baubles like Pepsi A.M.

Absurd baubles—failures and successes in equal measure—are the stock-in-trade of the New Product Showcase & Learning Center. At the time of my visit the NPS&LC was domiciled in an industrial park on the outskirts of downtown Ithaca, just one of a handful of drab, low-slung, aluminum-sided warehouses. (The NPS&LC recently evolved into NewProductWorks and moved to Ann Arbor, Michigan.)

It was a grey, drizzly day, archetypal in one of America's cloudiest enclaves. I was received by the NPS&LC's affable founder and director, Robert McMath, and his trend-mongering colleague, Marilyn Raymond. They ushered me quickly through the main display area—an enormous warehouse space with the layout and contents of a ludicrously overstocked supermarket—into a more business-like meeting room.

McMath was no slick, lingo-spouting corporate player. He came across as gracious, unassuming, often positively avuncular in demeanor. This, for example, was his explanation, in full, of the NPS&LC's genesis: "As a Scotsman, I never throw anything away."

Once I'd seen the place, I was ready to believe him. Piled throughout the building—on side tables, on shelves, stacked in corners, and jumbled and crowded onto floor-to-ceiling shelves in the centre's main warehouse—was the largest collection of consumer items I'd ever seen in one place. Imagine if, rather than regularly turning over its stock, your local supermarket kept at least one of every single thing it had sold since the 1970s on its shelves; this is what the NPS&LC felt like. It was a polyglot temple of consumption, a staggeringly detailed and multihued illustration of the sheer variety created by the consumer ethos.

It was, on the surface, an absurd place. It was not, however, a particularly corporate place. It was not polished to an ultraslick sheen. There was very little self-importance or self-aggrandizement evident in its signage. It did not look like it was trying to sell me anything.

I found this disconcerting. Wasn't I seated in the antechamber of a sacred crypt, mere steps away from the storehouse of some of the corporate world's rarest and most prized artifacts?

After the interview, McMath and Raymond walked me around the meeting room. It was lined with shelves, each overstuffed with some of the NPS&LC's current favourites, grouped into something akin to themes ("Co-marketing," "Soy & Ginger," "Caffeine"). I nodded knowingly as they talked of "nutraceuticals," made note of the recent rise in orange-flavoured products. There was talk of something called "cool fusion."

And, then—without the slightest trace of fanfare—there it was: a single can of Pepsi A.M. It was on a shelf labelled "Favourite Failures," tucked between a few jars of Gerber's short-lived "Singles" line of baby-food-style products for adults (sample meals: "chicken Madeira" and "beef with mushroom gravy") and a bottle of Richard Simmons "Salad Spray."

It was a standard can, 12 ounces in volume. The brand name and the red, white and blue yin-yang logo were rendered in the instantly familiar way against the can's white background. The only modification was the suffix "A.M."—done in a faux-handwritten script, at a slight angle, as if to suggest a certain rakishness. It was quite clearly a Pepsi product. Just another Pepsi product. It was real.

I asked McMath where it came from, working hard to keep from appearing deranged. I wanted to hold the can aloft, shout *Eureka*!, dash from the building with my precious artifact, collapse in a giggling fit. But I kept my cool as best I could and listened to McMath's telling of Pepsi A.M.'s tale.

The mythic version had the date right. Some time in the late 1980s, someone down PepsiCo way took note of the fact that some people (perhaps a growing number of them, enough even to indicate an emerging trend!) had taken to drinking cola at breakfast instead of coffee or tea. Why not, thought the PepsiCo people, create a Pepsi specifically for these early-morning cola drinkers? Why not, in short, create a breakfast cola? So they came up with a brave new Pepsi, higher in caffeine, lower in sugar, and not quite "Pepsi Cola" in flavour. (McMath could not confirm the myth's claims of lower carbonation.)

In 1989, it was brought to consumers in select locations … and it was never heard from again. McMath thinks Pepsi A.M. flopped because it didn't taste like Pepsi and because it introduced certain unwanted semantic problems into the consumer-cola relationship. That is, as McMath explained with some amusement, the consumer was faced with the quandary of what to do if it was past noon and he or she had only a can of Pepsi A.M. in the fridge. Could Pepsi A.M. be properly drunk in the P.M.? Was the serious cola drinker now required to purchase—and keep on hand—two kinds of Pepsi at all times? McMath suspects that the average consumer simply did not want these kinds of dilemmas introduced into the cola equation. I'm inclined to agree with him.

The myth had been confirmed; my work was largely done at the NPS&LC. I took a few photos of the Pepsi A.M. can for posterity and prepared to leave, asking McMath if he'd mind if I poked around in the main display area before I left. I thought perhaps I'd find some oddball products to round out my business-magazine piece.

So I wandered off through the NPS&LC's long aisles of dusty groceries, jotting down obvious oddities—Male Chauvinist "Awfully Arrogant" aftershave, for example. As I came around the end of an aisle, I found myself staring down a long row of shelves lined, end to end and floor to ceiling, with soft drinks. Hundreds upon hundreds of them, stretching as far as I could see, past the point where I could even differentiate between labels. It was a riotous blur of soda, stretching to infinity. And suddenly the anticlimax of my adventure had found a final surging peak, like the last gasp of the killer in a bad horror movie.

I had never seen such a vivid illustration of consumerism's intrinsic restlessness, its unwavering refusal to accept that anything is finished, complete, fine just the way it is, thanks. First, there was the seemingly boundless variety on display, the relentless reinventions and rechristenings of the wheel laid out before me. There were, of course, the mainstays of the industry, the colas and root beers and ginger ales, each of them represented by a dozen or 100 brands. And there were the unabashed novelties, the freak flavours: lemongrass, sour blue raspberry, strawberry-coconut. Each of these was intentionally curious, speaking only to the fact that nothing is too contrived to make it to market in the consumer age.

Beyond these were alarming attempts at further invention: a handful of fudge sodas; an equal number of bubble-gums; all manner of queer hybrids (blackberry-guava, almond creme, "Fruity-Nutty Cola Flavor"). There was one that claimed to be "double fudge." And there were no less than a dozen coffee sodas, including Pepsi's own Kona, attesting to the fact that Pepsi A.M. was a more mundane myth than I'd thought.

And then I took a closer look, started to dig deeper, to categorize and classify. The colas proved a particularly fertile area of inquiry. There were just so damn many of them, so many names and logos and personalities given to what was more or less the exact same drink. There were the requisite dozens of Pepsi and Coke products, as easily discernable as opposing armies, huddled together with their ubiquitous logos and telltale colour schemes. But there was also a veritable United Nations of petty despots, an entire global geography of colas. Here a can of Tianfu China Cola, solitary and aloof; there, a can of Mexicola, just down the shelf from an Inca Cola in tarnished gold. There were Faygo Caribbean Cola and Goya TropiCola Kola Champagne and Chubby Cola, all brightly decorated and covered in symbolic allusions to warm sun and fine sand. On a more defiant tip, AfroKola billed itself variously as "El Sabor de Libertad," "The Taste of Freedom" and "The Soul Drink." And there were of course homages to cola's birthplace: U.S. Cola and Cola in U.S.A. and Sam's American Choice Cola (though I suspect only the last of these was actually manufactured there).

The geography lesson completed, the colas then spelled out more subtle lessons in the cultural history of the American century. The lecture began and ended with Pepsi and Coke. Their older bottles and timeless logos spoke of a young nation's emergence as an industrial powerhouse, so full of energy and promise. More recent deviations—not just Pepsi A.M. but Pepsis XL and Max and Free and a dozen others, plus Coke's less numerous but more monumental mutations (especially New Coke)—suggested a certain smug, self-important carelessness, these long-forgotten "innovations" reminiscent of an imperial king's battlefield misadventures.

The NPS&LC'S colas told smaller stories as well. Sunday Funnies Cola, its label sporting the visages of Blondie and Beetle Bailey and

Hagar the Horrible, seemed to evoke nostalgia for a time before the rise of television, while Jazz Cola made reference to an edgier part of the same era. Dis-Go Champagne Cola spoke clearly—if only semi-literately—to the decadent and indulgent 1970s. And the 1990s were memorialized on the side of a can of O.K., A Carbonated "Beverage" [sic], a brilliantly succinct summary of the youth culture's irony-soaked rejection of commercialism and the co-optation of same by mass-marketeers.

This was just the colas. When other genres of soft drinks were included, the scope of the carbonated beverages became truly global. Shasta Yoshi Apple Soda added both Japan and the Pacific Northwest to my world map. First Nations Grape Soda and Wild West Firewater ("The Flaming Red Soft Drink of the Old West") told conflicting tales of the American frontier. Way 2 Cool All Natural Root Beer hinted at a particularly eighties sensibility, some twisted mix of hyper-confidence in the present and nostalgia for a simpler time. And Mad River's random mix of references—Black Cherry Explosion, Revelation Root Beer, Wicked Blackberry Guava—read like a concise summary of postmodernism's recent, rapid decline into nonsense.

I mean, surely a culture that gives rise to—that is, creates a market for—products like Love Potion No. 69 Arousing Carbonated Drink and Oop! Juice Carbonated Ghoulade and, for that matter, Pepsi A.M., must be in the grip of some very strange mania.

OF COURSE, CAPITALISM, ESPECIALLY in its post-World War II corporate incarnation, demands constant growth. Regardless of its achievements—and the ability of carbonated sugar-water to insinuate itself into the diet of nearly every society on the planet is an enormous one—it is not allowed merely to coast along. It is predatory. Sharklike. It must keep moving, keep feeding itself with new and new and new, or it will die.

Classical economic analysis, with its talk of sufficient supply meeting demonstrable demand, no longer applies. If there is no demand, it must be created. Supply must constantly be expanded. More food for the great beast, or it will soon be a corpse washed up on shore. Much as Nike's design team has, on occasion, churned out an entire new shoe design each day, so must the makers of carbonated sugar-water hunt constantly for novelty. A new flavour, a new colour, a new name that taps into a lifestyle trend. Anything. Yesterday is irrelevant; there must be more products—new products—on the shelves tomorrow.

Had Pepsi A.M. proven capable of making money, of keeping the company's bottom line growing, of keeping the shark moving through the water—then it would have been noteworthy. Then, perhaps, it would have made it into PepsiCo's official corporate history.

In PepsiCo's telling, the year of Pepsi A.M.'s test launch, 1989, was a busy year. PepsiCo began awarding stock options to its full-time employees. It acquired two top British snackfood companies and "the Smartfood ready-to-eat popcorn business." And it was named one of Fortune magazine's 10 Most Admired Corporations. But the debut of Pepsi A.M.? Irrelevant. A rumour, a myth. A legend. It might as well never have happened.

"Answering teenagers' call for a berry-flavored, blue-hued cola"— this just in, from a PepsiCo press release dated July 22, 2002. "PCNA [Pepsi-Cola North America] bottlers this week are beginning to deliver PepsiBlue to store shelves across the country." That's not all: "Nine months in the making, Pepsi Blue was one of more than 100 cola fusion concepts tested by PCNA's innovation and R&D groups." And this, from PCNA chief marketing officer Dave Burwick: "It looks and tastes unlike anything else on the market, and we believe it can redefine and grow the cola category, particularly among teens— the next generation of cola consumers."

In other words, it's something brand new. It's history in the making.

BUZZ, INC.

Shift, November/December 2000

One of the things Shift *did really well was explore an emerging phenomenon in depth without feeling obliged to predict its trajectory or pass judgement on its import. Too much technology writing morphs sloppily into futurism; it was more worthwhile, to my mind, merely to describe what the strange new rocket looked like in flight. This was one of those snapshot-of-a-rising-star pieces, looking at an early adopter in a viral marketing game that would eventually become as commonplace as a Google ad or corporate Twitter feed.*

I look forward to being a part of this.

If I can help in any way just let me know.

thanks for being so accessible. its very cool of you.

I would love to be involved with the band more in any way. Promotion is always cool. Just let me know what I can do.

I'm spreading the word my friend!!!

im putting more stuff up daily and it'll blow your socks off pretty soon !!!

thanks for mentioning me in the news flash! that was cool :)

I can't wait to get the sampler tapes so I can start passing them out!

I'm really sorry about how little I've done. This week has been really busy, but next week I will definitely have more time to promote.

Thanks for everything Alicia! You're totally cool!!!

I just got the tapes and matchbooks and stickers and TICKETS now!!!!!!!! THANK YOU SOOOOOOOOOOOOOOOOOO MUCH!!!!

THANK YOU THANK YOU THANK YOU!!!!!! You are the greatest person ever!!! I LOVE YOU!!!!!!!!!!!!!!!

MAN, THE THINGS KIDS ARE SAYING to music-industry marketing reps these days.

Then again, these aren't just any kids: They're members of an "online street team" doing voluntary promotional work for a band called Unified Theory. And the marketing rep to whom they're addressing these warm fuzzies isn't just any marketing rep: She's Alicia Dooley, a grassroots marketing coordinator at ElectricArtists. And those warm fuzzies? Totally heartfelt, near as I can tell. And the by-product of an approach that appears to be well on its way to reinventing the way cultural commodities are marketed.

But I'm getting ahead of myself.

Let's start with Alicia Dooley. She's twenty-two, a recent graduate in neuroscience and behaviour from Columbia University. She's also a rabid music fan, which helps explain why she's working at ElectricArtists.

As for ElectricArtists, it's an online marketing firm based in New York. Its clients include the requisite Telco giant (AT&T), a handful of movie studios and every major record label in existence. EA does quite a bit of interesting online promotional work: giving away an unreleased Alanis Morissette song, prebooking tickets to Depeche Mode concerts, selling Garbage's signature line of cosmetics and the skateboard Shirley Manson designed herself. The company also helped *Rolling Stone* sell truckloads of a recent issue featuring NSYNC by emailing a JPEG of the cover to the members of the band's various fan clubs. But where EA has really attracted attention is in what it calls "grassroots" or "community-based" marketing (a tactic that buzzword-mongers might recognize as a variation on the viral-marketing theme). If the name sounds familiar—ElectricArtists, that is—it's probably because the company is given a sizable share of the credit for turning Christina Aguilera into a pop superstar.

It was this last trick—creating that amorphous but extraordinarily powerful cultural force known as "buzz"—that brought EA the most acclaim (and laudatory business-press coverage). The campaign tied in with an offline promotional blitz, of course, but the critical component that EA contributed—which paved the way for the print ads, the slick videos, the MTV appearances—was the organization of Aguilera's fan base into an active, awareness-generating online community. There are two tricks to this, neither of them as easy as they might sound: (1) finding the potentially diehard fans; and (2) establishing a relationship with same. With the Aguilera initiative, it began with research into where her fan base was and what they were talking about. "The term is 'lurking,'" says Jason Madhosingh, who worked on the campaign. Which is to say, he and other EA staffers hung out silently in discussion groups, surfing teen-pop message boards and fan sites. And, eventually, joining the conversations. "It's much like going to a party," says Madhosingh. "You wouldn't walk up to a group of people who are already having a conversation and then say what you have to say and then walk away." Instead, EAs staff slowly and carefully nurtured relationships with the kids who would later set up Aguilera mailing lists and flood TV and radio request lines with demands for "Genie in a Bottle." We can get into the finer details of this delicate process later; for now, it's enough to note that Aguilera's label was very pleased with the results.

What EA did for Aguilera is roughly the same thing that Dooley and her team are now trying to do for Unified Theory, a band consisting of the three surviving members of Blind Melon (Chris Shinn, Christopher Thorn and Brad Smith), plus Pearl Jam's original drummer (Dave Krusen). Its self-titled debut came out on Universal in mid-August. When EA started working for Unified Theory, back in May, they found that the band had an existing fan base, but that it was somewhat scattered and dormant. What EA does next on such a marketing campaign is best explained by its effusive thirty-six-year-old CEO, Marc Schiller: "Our grassroots marketing staff, they literally organize these fan bases. They send out information to them on a weekly basis. They activate these fans to then go out and spread the word virally, however the fans want to do it. We don't give road maps or scripts. We don't say, 'Email thirty people, and if you email thirty people, you get a CD.' We don't do that, because that then becomes spam." Instead, EA simply provides its online street teams with raw materials—advance copies of albums and plenty of information—and allows the fans themselves to pass the stuff on to friends and net acquaintances. The volunteers are then rewarded for their work with concert tickets, souvenirs, more insider information or all of the above.

These, then, are the broad strokes. At online street-level it's a much more sophisticated process. But before we get into all that, let's take a walk out the door of EAs offices on West Forty-first Street in Manhattan. down to offline street-level. Let's go to the southeast corner of Forty-fifth Street and Seventh Avenue in Times Square, just a couple blocks away.

It's just another Wednesday afternoon in Times Square. Stand here at the curb and watch the cars go by in a dozen steady streams. The pedestrian traffic on every sidewalk is almost as thick and just as relentless. Take a look around. There's the Virgin Megastore—it's the one girded in storey-high pitches for Christina Aguilera, NSYNC, Eve 6 and others. Dig the surreal, visual Doppler effect created by those three scrolling bars of pixelboard gold as they bend around the corner of that Nasdaq-encrusted building over there. Look south, and read more odd-angled pixels hawking ABC, NBC, the business news, the weather. Check out the billboard advertising some sci-fi movie— that one's gotta be at least ten storeys—or the five-storey-high bowl of Nissin Foods Cup Noodles, or the fifty-foot-long Bud bottle, or the block-long neon URL www.feveronbroadway.com. A vast, relent-lessly churning sea of people and information. Not a bad analogy for the internet, is it?

Now, to get a better sense of what ElectricArtists does, take a look at the southwest corner of Forty-fifth and Seventh. There's an exuber-ant knot of about a hundred people there—predominantly but not exclusively teenagers—and they're cordoned off in the right-hand, southbound lane of Seventh, behind NYPD-blue sawhorses. With their necks all craned back at about the same angle, they're gathered there just to stare into the *TRL* aquarium, one storey above street level. (In case you're wondering: *TRL* is *Total Request Live*, MTV'S gargantuan hit of a daily top-ten countdown show. It broadcasts live from a studio that looks out, through three huge picture windows, onto this very intersection of Times Square.) A sizable minority have brought homemade signs pledging their devotion to assorted pop bands, or to the show's host, Carson Daly. (One kid, who's wearing a swimming cap, dive mask and snorkel, stands stoically behind a sign that reads: SNORKEL SIMPLY LOVES *TRL*.)

This gaggle of fans is responding in near-perfect unison to cues from the aquarium: If the girl in the headset up there holds her hands above her head and claps—pandemonium on Forty-fifth Street; if she drags her hand across her throat in a pantomime of cutting—instant silence; if Carson Daly comes to the window and peers out—pandemonium squared. Meanwhile, a half-dozen attractive young women in form fitting, shiny red vinyl pants and white, logoed tank tops work their way through the crowd, handing out T-shirts at random to the assembled gawkers. The T-shirts, logoed the same as the tank tops, read: "Emazing. com—Whatever you're into, we email it to you. Free!"

But what if, instead of simply doling out T-shirts indiscriminately, one of those vinyl-panted babes actually noticed, say, that one kid in the crowd was wearing a Pearl Jam T-shirt? She could approach that kid, talk to him, get to know him a little, and then hand him an advance copy of Unified Theory's new album and a newsletter full of information about the band. Maybe she'd say, "Look, just give it a listen, and if you like it, maybe you'll tell your friends about it." And if the kid does all that, he can get back in touch with this girl, and then instead of just catching a glimpse of his new favourite band through the picture windows on Times Square one day, he gets a ticket to their next concert, or maybe even a signed poster. This is pretty much what ElectricArtists does.

ALICIA DOOLEY FAVOURS CASUAL, dark-hued clothing—not red vinyl—and doesn't seem like the kind of person who'd be very comfortable accosting people in Times Square. She is, however, right at home trading email and Instant Messenger messages with Unified Theory fans. Most of her day is spent in a cramped corner office at EA, where she and a handful of other grassroots marketers sit amid a 3-D patchwork quilt of music-industry freebies (the kind of stuff they reward their street-team members with). Right now, though, she, Marc Schiller, Mary Reilly—who, at twenty-four, is a "grass-roots/community marketing manager," and thus is sort of Dooley's supervisor—and I have moved down the hall to a gutted office that will soon house some of EA's rapidly expanding staff.

Schiller has just finished one of his long and invariably captivating rambles, this one about how the key to EA's success is that they've learned to walk the line between information dissemination (which young people generally like) and advertising (which young people variously distrust, ignore, mock and outright detest). He was, in fact, just explaining that EA staffers have made it their unwavering policy to

always—"under all circumstances"—identify themselves as marketers when they interact online with fans. (In October 1999, the *Wall
Street Journal* reported information to the contrary. While researching
the story, the *Journal* reporter had asked Schiller if EA staffers always
disclosed their identities. She had, Schiller said, underlined "*always*."
Wanting to be sure of his response, he asked one of his employees about
it. The staffer admitted that on one occasion during the Christina
Aguilera campaign he'd wanted to get honest opinions about her new
video, so he'd gone into a chatroom and asked what people thought
about it, identifying himself only by his first name. Schiller—who is
one of the must unassuming, forthcoming marketing guys you could
ever hope to meet—duly passed on this information to the reporter. The
Journal article then made it sound as though this was standard operating procedure at EA. Schiller knows better than that. "The second that
the fans feel it's not authentic," he concludes, "is the day that not only
the marketing stops, but the whole credibility of the artist stops.")

Schiller dashes off, and then Dooley explains to me how, in practice, people like her—whose name is the one next to the company's
on the street-team online newsletters she writes—earn authenticity
points: "Like, I'll be in Napster or something and do a search on
Unified Theory. So I'll just Instant Message [someone who has posted
a bunch of Unified Theory songs] and say, 'Hey. I noticed you're a
fan …' And sometimes it will be a street-team member, and they're
like, 'Wow, you're really Alicia who writes the newsletters?' And I'll
talk to them. And they're like, 'It's really cool to talk to you, and realize that it's a real person doing this.'"

"I think," Mary Reilly adds, "that we've certainly changed a lot of
their opinions about what marketers do. We've recruited volunteers
from various types of online sources, who are like, 'Wow, you guys are
marketing? I mean, it sounds like your job is so cool.' I'm like, 'Yeah,
it is kinda cool.'"

Dooley: "Yeah, they realize that marketing doesn't have to be a
bad thing."

Reilly: "Yeah. It's not, like, 'The Man,' you know?"

Dooley: "It's great. It makes your day."

Yes, they are marketers. They love their jobs. And, most importantly, they're good at them. And they're good at them because they

actually *are* music lovers, pup-culture junkies, frequent Instant Messenger users—*fans*—and so all those qualities that the mass-media advertisers of the world try so hard to don as if they were Halloween masks—authenticity, respect, youthful enthusiasm—come naturally to them. And they're also good at their jobs because they're allowed simply to be themselves. EA recognizes that when today's youth reject advertising, they are not—a small subculture of hardcore culture-jammers excepted—rejecting the entire consumer ethos. They still want to own the CD, see the show, buy the T-shirt. They just don't want to find out about it from a dumb TV commercial or an even dumber bulk email. They want to hear about it from a real person.

A short while later, Schiller returns, just as Mary Reilly is explaining that she likes her job because EA doesn't produce transparent, inauthentic ads: "It's more about building a community."

Tricky word, that. It's a word that Schiller and his grassroots staff use frequently, because building community is, in fact, what they endeavour to do. Now, I hear that, and my reflexive response is: Marketing companies cannot build real communities. I mean, marketing is, at its core, about persuasion and manipulation and nasty stuff like that, right?

But then I consider that Unified Theory's official website has a link to a message board, and that about twenty fans use it regularly. They hang out there. Here's a representative thread, entitled "Hey—what's up!!!!????"

> Yodulf: Did everyone go on fricken holidays, or something!
> Evolution: As a matter of fact I did! [smiley face]
> nothinglikeyou: You certainly did! [winking smiley face] Hee hee! [smiley face] How's Cali?
> Evolution: It's awesome here! I'm having a lot of fun … fortunately all my friends have computers so I can keep up with things!!! [smiley face]
> nothinglikeyou: Then you get to stay in the loop, oh Sweetness and Light! [winking smiley face]

Looks about the same as any other online community, I figure. Then I learn that this message board was set up by a member of EAs online street team, and I'm forced to re-evaluate. I can even see why some-one—a would-be competitor, say—might get to thinking that this community-building that EA does looks like pretty simple stuff, not to mention awfully effective.

Back at EA's offices, Schiller addresses this misconception. "You could say, 'That's pretty easy to do, man. Getting fans together and talking about things—I could do that.' A company could say, 'Well, let's just do that in-house.' Trust me: once they try to do it, once they try to communicate like Alicia communicates ..." He trails off, letting me fill in the blank: They fail. "A lot of these companies, I know, try to do it themselves. And what comes back is a massive amount of negativity, because they didn't understand it. There's no way they can navigate these waters. I mean, there's nothing more perilous than ten thousand teenage girls. Nothing." Schiller laughs. And to sail these seas takes time—Reilly, for example, spent about eight months in close contact with Red Hot Chili Peppers fans for a recent campaign.

As well, EA has a substantial head start on the competition. (The only other marketing company ever mentioned alongside EA is Los Angeles–based M80 Interactive.) A quick history: Circa 1994, Schiller was at House of Blues, heading up their infant new-media wing; his pal, EA president Ken Krasner, had much the same gig at RCA, where he oversaw the net's first-ever major-label album pre-sale. They started talking about the power of this new medium. Krasner jumped first, founding his own company—Internet Music Marketing—out of his living room, in 1996. He started to build a client list. Then Schiller joined, and, in 1997, EA was born. Two years later came the Christina Aguilera campaign. And then things really started taking off. These days, EA is "a multimillion-dollar company" (privately financed by its founders, so they don't proffer hard figures). Not only a fair chunk of the entertainment industry but even reps from such decidedly un-teen-pop fields as financial services and phar-maceuticals have come calling.

"Christina Aguilera," Schiller says, "was our tipping point."

AND SO WE COME—INEVITABLY, REALLY—to Malcolm Gladwell's recent book, *The Tipping Point*. "When I read it," Schiller tells me (it is just he and I now, in EA's boardroom), "it was as if he'd been part of our conversations. And really, I think he's definitely on point."

Gladwell's argument is that social interactions follow the same basic rules as those that govern the spread of viruses, of epidemics. That is, when a social trend—an idea, a product, a message—moves through the culture, a very small proportion (say, between five and twenty percent) of those "infected" do most of the "infecting" (at least eighty percent). The former group is made up of a mix of people Gladwell calls Connectors (people who have way more friends than most), Mavens (people who know things before anyone else, and share this information) and Salesmen (people who are particularly gifted at making a radical idea/product/message seem attractive to the majority). Gladwell's book isn't exclusively—or even primarily—a marketing text, but if someone were to set up a marketing company based on his ideas, it'd probably look a lot like ElectricArtists.

This is not simply because EA's staff and its online street teams undoubtedly contain their fair share of Gladwell's Mavens and Connectors and Salesmen. Equally important is the fact that his argument—that you don't have to pass on your message to everyone all at once (à la traditional mass-media marketing), you just have to get it to the right tiny percentage—is the principle underlying EA's grassroots campaigns.

"If you're a fan," Schiller says, "you are marketing that band. You're marketing the band in the T-shirt you wear, you're marketing the band in the fact that you have to have the latest record. So if we go to that person, we know that that person is the tastemaker and is going out there and drumming up an authentic support for that band." All EA needs to do is find such people—a major component of the jobs of both Mary Reilly and Alicia Dooley—feed them information and reward them (on the Unified Theory campaign, for example, street-team members have received advance CDs and CD singles, signed album art, stickers, autographed posters, T-shirts and tickets).

This kind of marketing is, of course, nothing new. It dates at least as far back as the Who's 1965 London club gigs, when the band's managers turned their hardcore fans into a club called "The 100 Faces," who put up posters in exchange for tickets. More recently, street teams—referred to as "offline street teams" at EA—have been assembled to poster and spray-paint city streets for a wide variety of artists (particularly hip-hop acts). But the *power* of this kind of promotion, when it's done online, *is* new. A posting on the right message board is like slapping a poster up on a street-corner lamppost that stands, miraculously, in a place where only those already interested in the message walk by. And, moreover, where the poster looks not like an ad, but like a friend's sincere recommendation.

HERE'S SOMETHING THAT HAPPENED to me while I was hanging out at EA's offices. I can't seem to fit it in anywhere else, but I'm sure it says something about the company. It goes like this: I was in the boardroom with Mary Reilly and Jason Madhosingh when Marc Schiller's assistant, a pleasant young man named Eric, appeared at the door.

"Sorry, guys," Eric said, "but we really need the room now."

Then there was a gruff voice, mock-angry, from the hall. It all happened very quickly, but the voice said something like, "Yeah, what's wrong with you kids? C'mon, get outta here!"

I quickly piled my notebook and tape recorder into my bag, turned to leave...and found myself face to face with: Dick Clark. America's Oldest Teenager.

Perhaps the circle was closing. Perhaps a torch was being passed. Something.

I HAVE NOT TALKED MUCH ABOUT Ken Krasner, thirty-nine, EA's president and cofounder. This is primarily because, for most of my time at EA, he was a shadowy figure, zipping in and out of this or that office with a quick handslap or "How's it goin'?" He was kind of like Cancer Man from *The X-Files*, if Cancer Man were a nonsmoking, high-spirited, avuncular sort dressed in a blazer and jeans. Late in the afternoon of my last day there, though, he waved me into his office— with a deft combination of conspiratorial slyness and whimsy—and we sat down to chat. Off the record. "No, forget the notebook," he said when I reached for it.

And what sort of "deep background" stuff was revealed? Well, Krasner wanted to make sure my visit was going well—that I was getting the information I needed, that I'd met with everyone I wanted to, that sort of thing. He also divulged that the guy playing the conga drums stage-right in the photo of a 1980 Gloria Gaynor concert that he had on his wall was him.

(To make sense of this next part, I should mention that there are quite a few Indian trinkets scattered around the EA office—a Ganesh figure on the receptionist's desk, a couple of paintings of the Buddha, a Sai Baba photo on Schiller's desk—and that Schiller, for one, has been to India several times. And that none of this came across as mere dabbling in Eastern spirituality. It didn't come across like the office was peopled by gurus, either, mind you; it just seemed, I don't know, genuine.)

Our meeting coming to an end, Krasner leaned in across his desk. "Our karma is clean," he told me. He paused, leaned back. "I'm happy about that."

Know what? He meant it.

Know what else? I believed him.

Then we went out into the hall, and he and Schiller made cracks about who'd put on more weight recently.

THE DIRT ON THE SMOKING GUN

Shift, December 2002

Reporting on journalists—especially ones as seasoned and observant as the Smoking Gun crew—is a delicate business. They know your tricks. They understand exactly what kind of detail you're looking for. They don't forget you're in the room. They're onto you. Still, I remember the reporting for this piece so fondly that I pretty much forgot it consisted almost entirely of sitting as unobtrusively as possible in a single Manhattan office space for a few days. The whole wide tabloid world of celebrity infamy seemed to be in the room with us. As ever, it paid huge journalistic dividends to actually be there.

After the story came out, the Smoking Gun's Daniel Green paid me the ultimate journalism pro's compliment—he wrote to say mine was the first story about their enterprise that didn't make a single factual error.

I. THE CURIOUS CASE OF JOHN GOTTI'S DIRTY LAUNDRY

THE FIRST THING ABOUT THE old laundry of recently deceased New York Mafia kingpin John Gotti is that it's pretty nondescript. The second thing is that if you find yourself standing in an equally non descript office on the seventh floor of a quite similarly nondescript building in midtown Manhattan with some of it in your hands—say, for example, the white mesh laundry bag that John Gotti used to transport his prison uniform and other linens from laundry room to prison cell at the United States Penitentiary in Marion, Illinois—if you find yourself in such a situation, trust me, it's spooky in a distinct sort of way that's hard to explain.

There was a voyeuristic twinge to it, to be sure, a clear sense that I was holding something that I should not have been holding. I wondered, dumbly, whether it was illegal to be in possession of the laundry bag of a deceased organized crime boss. Or whether Gotti's goons could, you know, *get* me for being part of it. And there was, as well, something almost subversive about this spookiness. Right there in my hands was the laundry bag of a man who, not wanting his glamorous image tainted by the pedestrian shame of his final days, managed to keep the number of known photographs of himself in a prison uniform to a grand total of one. And finally there was a feeling of certainty, a sort of momentousness, in much the same way that the sight of an empty field somehow becomes informative and powerful and moving when it is seen over the top of a plaque that reads, for example, "The Battle of Waterloo." And so there was a white iron-on tag on the laundry bag—"Gotti, John / Bin A-94 / 18261-053"—and a sense that this was something that should be looked at, was worth looking at, needed to be seen ... but not for too long.

The Smoking Gun (TSG)—the website whose office I was standing in when I had my little moment with John Gotti's laundry bag—specializes in finding and disseminating these sorts of items, in allowing the general public to peruse documentary evidence that would otherwise remain in obscurity. The Gotti uniform was to be their next big scoop.

It'd been pretty quiet at the offices of TSG for most of that day, just a steamy, sleepy July Friday in New York, the room quiet save for the scabrous squawk of Fox News in the background. The "offices" of The Smoking Gun consist of a series of desks arranged in a semicircle along the walls of a single alcove the size of a suburban living room. The alcove is at the end of a long hall, an ignored corner of a warren of offices occupied by Court TV, the cable network that owns TSG. Three of TSG's four staff members—co-founder and managing editor Daniel Green and reporters Andrew Goldberg and Joseph Jesselli— were tapping away at keyboards or talking quietly into phones at various spots around the room.

Enter William Bastone, the site's other founder, editor-in-chief and acknowledged driving force. With his short dark hair flecked with grey, his cuffed jeans and canvas Converses, Bastone looked more like a suburban dad in town for a Yankees game than an investigative reporter back from a quick lunch, an impression that faded as the electricity level in the room—as is typical when Bastone enters— seemed to at least double. Suddenly there was activity, banter, action.

At the desk nearest the entryway, Jesselli called out, "There's a fax on your desk." His broad, slightly glottal Bronx accent traced an arc halfway across the room's empty central space, the first stretch of a vocal Triborough Bridge.

"This is too good," Bastone responded gleefully, his Queens drawl finishing the trip across the East River. "This is too fucking good." Bastone was staring down at a sheaf of papers in his hands as he spoke, his head shaking to and fro in joyful disbelief, his grin even broader than his accent. The papers in his hands included the work reports of one John

Walters, prepared by his supervisors during his tenure as a labourer in the laundry room of the United States Penitentiary in Marion, Illinois. They confirmed: 1) that Walters had worked in the penitentiary's laundry room; 2) that Walters' performance had been "outstanding"; 3) that Walters had worked in the laundry room at the same time that New York Mafia kingpin John Gotti had been an inmate at Marion; and therefore, and most importantly, 4) that the prison uniform, laundry bag and handwritten note Walters had previously sent to the office were in fact the prison uniform, laundry bag and handwritten note of Marion prisoner 18261-053, the recently deceased John Gotti.

Bastone dropped the papers on his desk and crossed the room to a shelf near the entryway, bent over, and produced a large FedEx package. The TSG guys traded quips about Walters and Gotti as Bastone futzed with the package, eventually producing a pair of khaki trousers, a matching shirt and a mesh bag. Photos of these items would be posted on the TSG site (thesmokinggun.com) the following Monday—along with the strange, elaborate story of how Walters spirited them out of Marion upon his parole and attempted to sell them on eBay, how TSG reporter Jesselli happened to be surfing the site a few days after Gotti's death looking for weird Gotti memorabilia, how Walters inexplicably identified himself by name in a subsequent fax, and how the TSG guys therefore obtained and confirmed the authenticity of the garments. A full-page story about the clothes would also appear in the *New York Post* and a couple of British tabloids. (The *Post* acknowledged its debt to TSG; the British tabs did not.)

For now, though, they were right here in the room. Bastone handed me the bag—what I presume to be a standard-issue federal-penitentiary laundry bag—and I inspected it. Something about it made my stomach jump just a little.

"I had those things in my apartment for like four days while I was doing the pictures," Andrew Goldberg told me just then, his accent trucking down the New York Thruway from somewhere not too far upstate. "It was kind of weird, having this mass murderer's clothes in my apartment."

No fuckin' doubt. And it's that weirdness, the visceral thrill of seeing something you think maybe you shouldn't be allowed to see, the almost-illicit kick that comes from rifling through someone else's dirty laundry (in this case quite literally)—that feeling is one of the main reasons why, amid the rubble of hundreds of content-driven websites, The Smoking Gun has soldiered on. Once a day (roughly), every day (more or less), TSG posts something like this: a document, a mugshot, a transcript, some piece of raw meat from the vast slaughterhouse of the day's news. Here is Jeb Bush's daughter's arrest report. Here is Nick Nolte's mugshot. Here is the INS document approving an extension on Mohammed Atta's student visa, received by his former flight school in March 2002. Take a look, inspect it at your leisure, and decide for yourself what truths it brings to light. It is a simple, straightforward idea, it is inexpensive to put together, and it is pulled off with skill and aplomb by the TSG guys. Which is why The Smoking Gun has to be counted as one of the most successful websites of the internet's first generation.

II. WEB PUBLISHING 450: A CASE STUDY IN ADVANCED NICHE DEVELOPMENT

IN THE EARLY DAYS OF ANY new medium, there is an inevitable period of crude experimentation. In time, whether through theoretical breakthroughs or further technological advancements or sheer instinct, the new medium's most salient properties will become evident. The addition of sound to film catapulted it from novelty to mass medium to art form; television eventually learned to stop being filmed vaudeville; and so on. And thus are the producers of online content slowly learning that the web is neither an onscreen newspaper nor a five-billion-channel television set, but rather its own medium with its own strengths and weaknesses.

The case of The Smoking Gun is a case of that aforementioned instinct more than anything else. Neither William Bastone (now 41 years old) nor Daniel Green (now 39) were particularly proficient in computer programming—Bastone was a crime reporter for *The Village Voice* and Green was a freelance magazine writer—and neither knew much about the net when they decided, in early 1996, to enter the world of online publishing. As reporters, both of them—especially Bastone, whose specialty was Mafia stories—had collected considerable stocks of documentation in their reportorial endeavours: court transcripts, affidavits, FBI reports, that sort of thing. They thought it might be fun to publish some of it. They spent a year mulling over the idea and consulting experienced web publishers (for example, the proprietors of Time.com) about what strategies worked online. They were told to try to figure out who their audience was and what sort of site that audience would want. To their credit, they ignored the vast majority of this advice.

Instead, they followed their own hunch: that these documents they found so fascinating would prove equally so to the general public. They got Bastone's wife—Barbara Glauber, a graphic designer—to design the site and enlisted one of her former students, Mike Essl, to handle the technical side of it. On April 17, 1997, TSG went live with a refreshingly stripped-down look that perfectly underscored the site's just-the-facts approach and an FBI report documenting Elvis Presley's drug use. In subsequent years, it has added the occasional streaming audio or video file (a series of 911 calls made by a pointer named Willie that inadvertently got his master busted for marijuana trafficking, for example). It has also added a wildly popular special section called "Backstage Pass," consisting of the concert riders of more than a hundred musicians and bands—everyone from Sinatra to Ozzy to Britney to J-Lo. (Riders are the legal documents by which a touring performing artist requests food, refreshments and other personal items for the backstage area of the venue.) And it has spawned a single spinoff, WMOB.com, a wiretap "radio station" which publishes one episode per week of "The Frank & Fritzy Show." The show consists of recordings of casual conversations between Genovese crime family soldier Federico "Fritzy" Giovanelli and his cronies (particularly Frank "Frankie California" Condo), recorded in the mid-1980s; it is a character-driven show, equal parts funny and fascinating, much like *The Sopranos*. Beyond these small tweaks and additions, however, TSG has not changed in any substantial way in its five years online, despite being purchased for an undisclosed amount by Court TV in 2000; despite the fact that the Court TV buyout allowed Bastone and Green to quit their day jobs, devote themselves full-time to the site, and hire two new reporters; and despite the fact that the online battlefield around them became ever more littered with the corpses of other ambitious content sites.

The Smoking Gun's success is considerable: It attracts more than one million unique visitors each month. It is one of the few internet-only sites that is regularly used as a source (and sometimes even as the basis for entire stories) by the offline news media, and Bastone and Green appear regularly on TV shows like *Entertainment Tonight* and on countless radio shows. As well, TSG has spawned a book of the same title which has sold more than 30,000 copies, and there is talk of a TV series or special, either for Court TV itself or produced by Court TV for some other outlet. This success is thanks in no small part to the single-mindedness of its devotion. It has remained where it was born, seated comfortably at the intersection of several significant vectors in turn-of-the-century pop culture.

To wit: The web has emerged as a popular and powerful tool for obtaining additional information and deeper context about a given subject; this is what TSG adds to your average news story. Online current affairs sites can publish details about a news story in a volume impossible to duplicate in any other medium; this is another role TSG ably fills. A serious knock against online current-affairs sites is that their information is unreliable, rumour-mongering, inherently suspect; because TSG publishes only unedited, incontrovertible documents, it neatly and completely dodges this criticism. Finally, the offline media content inspires its own set of suspicions, derived as it often is from carefully constructed press-conference soundbites and meticulously planned PR events and cozy conversations between inside sources and insider journalists. The Smoking Gun blows holes through all this puffery and punditry and PR spin like—well, like a bullet from a smoking gun. "We don't want to do an interview with you," says Bastone. "We don't want to go to your movie. We don't want to put you on the cover of our publication. So we have none of those constraints, which gives us a degree of freedom that maybe other people may not think they have."

III. RICK ROCKWELL'S MOST SIGNIFICANT ACHIEVEMENT

BY EARLY 2000, THE SMOKING GUN was a well-known stop on many a netizen's morning surf. It'd even begun popping up in the much bigger sea of the pop culture at large from time to time. When the site published FBI documents in June 1999 revealing Timothy Leary's willingness to snitch on his counterculture peers, for example, its name appeared in major newspapers worldwide. But The Smoking Gun's full-blown mainstream-media debutante ball came courtesy of Rick Rockwell.

Yes, that Rick Rockwell. Rick Rockwell, the reality TV pioneer. Rick Rockwell, the star of *Who Wants to Marry a Multi-Millionaire?* Rick Rockwell, the first-class creep whose ex-fiancée's nine-year-old restraining order against him was dug up by the TSG guys just a few days before Fox was to re-broadcast the show, which revelation led them to cancel said re-broadcast as well as plans for a sequel.

It started, as many of The Smoking Gun's best stories do, with the kind of idle speculation which cliché-inspiringly often takes place in hair salons, at water coolers, or in locker rooms.

Daniel Green: "Bill [William Bastone] and I never watched the show, but the next day, it was on the front page of *The New York Times*. I played basketball that night, and I was having dinner after the game with a friend of mine who was on my team, and he said, 'You should have a look at this guy Rick Rockwell and find out if there's anything sleazy in his past.'

"'Yeah, yeah—that's a good idea.'

"The next morning I checked it out, and I saw one case, in Los Angeles, that a Rick Rockwell had been in a domestic lawsuit. I didn't know if he'd been married and divorced or what the deal was. So I called up and got a copy of the case file sent to me, and that's when I found out what it was. I still couldn't believe it, 'cause I figured that every entertainment writer in Los Angeles was trying to dig something up on Rockwell. And maybe they weren't. They just, they didn't find it."

In short order, TSG, too, was in *The New York Times* and seemingly every other paper and magazine and news broadcast in the country. Green again: "We had done things before that were pretty big, but the Rockwell thing just was, was huge. It was unbelievable."

Soon after, Bastone quit his reporting job at *The Village Voice*, and by the end of the year they had agreed to sell the site to Court TV. Thus did Rick Rockwell help, however inadvertently, to allow the TSG guys to devote themselves full-time to the project, which might well be the only positive contribution Rick Rockwell has ever made to our culture.

IV. TWO STORIES THE TSG GUYS LIKE MORE THAN THE "ROCKWELL THING," WHICH MAY GIVE YOU A CLEARER SENSE OF THE KIND OF GUYS THEY ARE

1. "Ben Affleck, Hollywood Hypocrite"

This is a personal favourite of Andrew Goldberg, who at 31 is the youngest of the TSG guys. The story traces its origins to the 2000 election campaign, during which Ben Affleck stumped vocally and actively for Al Gore. Affleck later speculated to GQ magazine that he might run for office himself one day. Goldberg, remembering fondly: "I remember Danny sittin' in the office, and he says, 'I bet you he doesn't vote.'" After an exhaustive search of voting records in every conceivable location where Affleck could have registered—a search which included a lengthy, pleasant conversation with Affleck's first Los Angeles landlord—Goldberg and his colleagues determined that Affleck had not registered to vote since 1992.

Goldberg: "It's serious reporting going into figuring out how to do something like that."

Jesselli [TSG's other reporter]: "But again debunking, and making someone, you know, [reveal] who they are. I mean here's a guy saying, 'It's very important to vote, it's very important to vote, it's very important to vote.'"

Goldberg: "And if he were a smarter guy he'd have registered by now, which he still hasn't done." [laughs hard] "Because I checked." [harder still]

Me: "All of which lends credence to what you were trying to prove initially, which is that Ben Affleck is full of shit."

Goldberg: "Right. Exactly."

Jesselli: "We didn't say that."

Me: "No. I just did." [all laugh]

2. "The Malcolm X files" and "Malcolm X Auction: Hot Stuff"

William Bastone's co-workers will testify, frequently and without prompting, to his genius. They speak with genuine gratitude about how much he has taught them about reporting, and they talk at length about his almost supernatural ability to scan a legal document at high speed and immediately pinpoint the single line that makes it interesting. I have seen him do it and it is impressive. The story of Malcolm X's red diary is one of Bastone's favourite TSG stories, and it is perhaps the most vivid illustration of Bastone's seemingly instinctive eye for detail, and of how The Smoking Gun is, in its way, contributing to the course of modern history.

Bastone, a passionate lover of documentary evidence, got the notion to go hunting through the vast files of the New York City Municipal Archives looking for fodder for the site. The archives are not properly indexed, merely arranged by date. So Bastone tried to think of famous New York criminal cases, and thought of Malcolm X's assassination in 1965. The file on the case filled three cardboard boxes. Bastone gets himself wound up describing this discovery, in much the same way an obsessive music fan might get wound up describing the recent purchase of an obscure bootleg record.

Bastone: "There was this unbelievable collection of material. It still amazes me that it was sitting in [the archives'] warehouse in Brooklyn. It was all the original police reports on Malcolm X's murder—the original grand jury testimony of everyone who saw it, interviews with the wife, photos. They actually had evidence envelopes that had shrapnel that they took out of his chin at the hospital, and out of his arm. Fragments from shotguns. They had the fucking shotgun casings that they took off the floor of the Audubon Ballroom. They're all in this fucking box."

There was also an empty evidence envelope, with a property clerk's voucher attached that read, "1965 RED DIARY WITH THREE BULLET HOLES." The envelope was empty. Bastone included the voucher in the large package of documents posted at TSG as "The Malcolm X Files."

Bastone again: "Fast-forward like nine months. I'm reading Page Six in the *New York Post* one day, and there's a little item that says, you know, Butterfields auction house in San Francisco will be auctioning the red diary that Malcolm X had in his pocket on the day that he was murdered. There's a bullet hole in it and it's specked with blood. And I went, 'What the fuck is that? What the fuck is this thing?' ... Long story short: We write about it, it gets picked up in other papers, they stop the auction and they withdraw the item. The family of Malcolm X finds out about it—they see it on our site—they start raising a stink about it." In due course, Malcolm's red diary was returned to his family.

V. SEVEN DETAILS ABOUT THE TSG GUYS THAT PERHAPS GIVE YOU AN EVEN CLEARER SENSE OF WHAT THEY'RE ON ABOUT

1. THEY SOMETIMES USE THE acronym FOIA—short for Freedom of Information Act—as a verb, as in, "So we FOIA'd them for the documents."

2. They were all appalled—really, honestly appalled—that TV Guide included Howard Stern's TV show among its fifty worst TV shows of all time.

3. There is a low filing cabinet in the centre of their office that serves as a sort of coffee table, and on a typical day it is covered in the following items: the *New York Post*, *The New York Daily News*, *Newsday*, *The New York Times*, *Us*, *People*, and the weekly supermarket tabloids *The Globe*, *The Star* and *The National Enquirer*.

4. After seeing a photo shoot in *The Star* in which Bruce Willis was wearing a green baseball cap with a John Deere logo on it, Goldberg spent twenty minutes singing the praises of John Deere ball caps. He then ordered one from Deere.com.

5. William Bastone's workspace is almost completely bereft of personal items, with the exception of a framed photograph of what looks to be a group of nine senior citizens relaxing under a tree on the grounds of a retirement home. It is, in fact, a photo of a group of notorious Mafia leaders, and it was taken at a federal prison.

6. The three websites—besides TSG itself—that are more or less permanently open on at least one desktop in the office are: The Drudge Report (the web's most famous political gossip column), Fark.com (a clearinghouse for weird news), and Socalnews.com (a subscription-only online wire service specializing in news from southern California).

7. On the day George W. Bush gave his second speech about America's corporate meltdown (the one where he stood in front of a banner reading "Strengthening Our Economy"), both CNN and CNBC were running the speech split-screened with a chart showing the movements of the Dow Jones and Nasdaq Composite indices as he spoke. The TSG guys thought that was absolutely fascinating. So did I, incidentally.

VI. THE WORD FROM CORPORATE

GALEN JONES, GENERAL MANAGER, Court TV Online: "We are looking at how we can sort of package their stuff internally for the network. That's sort of actively underway at the moment... We're looking at whether there's a good opportunity for us to do a special show with them. It would be like a Smoking Gun special."

Strangely, this appears to be the full extent of the parent company's current plans. The TSG guys claim rarely to see their corporate bosses and even less rarely (which is to say never) to be told what to do with their site. About the only evidence that anything is expected from their arrangement is the TSG guys' obsession with having their site cited, in print, as "Court TV's The Smoking Gun."

VII. THE POPSICLE FORMERLY KNOWN AS TEDDY BALLGAME

FOLLOWING TUESDAY. EARLY AFTERNOON. On the small TV on the windowsill in The Smoking Gun office, there's a CNN reporter standing in front of a Florida courthouse. He's got a document in his hands, and he's describing it in some detail, and it's giving William Bastone an apoplectic fit.

"You gotta get that woman on the phone," he barks at Andrew Goldberg.

Goldberg protests that he's been calling her every few minutes.

Bastone is unmollified. "She just needs to know that we need it immediately."

Moments later, Goldberg is talking to "that woman," a Floridian paralegal of his very recent acquaintance whom he has sweet-talked into trekking over to the Florida courthouse in front of which the CNN guy is posing, in order to get a hold of the document he's posing with and fax it to the TSG guys. The document is the last will and testament of Theodore S. Williams, a.k.a. the Splendid Splinter, the last man to bat .400 in a full season of Major League Baseball and the first professional baseball player ever to be cryogenically frozen by his son in the interest of one day selling his genetic material for profit. The Ted Williams will is The Smoking Gun's kind of document, and the TSG guys have been working obsessively all day to be the first news organization to post it online.

Goldberg's on the phone to the paralegal, Daniel Green's on his phone trying another route, Bastone's pacing around like a nervous manager in the bottom of the ninth with three on and two out, and Jesselli—who is a veritable encyclopedia of baseball trivia—is trying to remember what Ted Williams' other nickname was. (It was, he will soon recall, "Teddy Ballgame," which nickname will be used by the TSG guys a couple days later in the headline for their exclusive story of how Ted Williams' creepy son, John Henry, trademarked his father's famous moniker mere weeks before Teddy Ballgame became the late, great, cryogenically preserved version of same. And it was Jesselli's initiative, incidentally, that led to the scoop; in the kind of oblique-angled thinking for which The Smoking Gun should be famous, it occurred to Jesselli that if John Henry was such a sleaze-driving man, he might've thought to trademark his soon-to-be-iced pop's John Hancock. But I digress, and in an ever-more-Walter-Winchell-esque style.)

Long story short: They got the will from the lawyer-like lady and slapped that ballplayer's bequest up on their digital shingle faster than an Enron exec-sec at a shredder, if you catch my drift. "This is a good day," blabbed Bastone to his Bronx bud Jesselli once the fast-fax fever had cooled. "We got Frankenstein and Ted Williams on the front page of the site in a two-hour period." Not to mention, Bastone speculated, that they most likely beat the snooty scribes at such august addresses as Ap.org and Nytimes.com to the job. Not too shabby for a quaint quartet from the outer boroughs.

Later Bastone told this scribbler, strictly on the Q.T., that this was just the kind of scoopage of which his site is in seekage. "We don't really, we frankly don't look at other kinds of internet sites and go, 'That's our competition.' … We kind of think like, we want to beat the AP. We want to beat the *New York Post*. We want to beat *People*."

You heard it here first.

THE SIMPSONS GENERATION

Shift, September/October 2002

I'm paraphrasing via hazy memory, but I believe Shift *editor Neil Morton assigned this one by saying they needed something to tie together the retrospective ephemera in the tenth anniversary issue, a big wide narrative arc that told the story of "ten years in the life of the culture." Sure. And while I'm at it, why don't I rewrite the Bible as a piece of postcard fiction?*

I rarely start a piece with as little sense of where it'll end up as I did with this one. The only breadcrumb trail I had to follow was my confidence in the assertion that there was a Simpsons *quote for every occasion. In retrospect, it seems especially fitting that I used* The Simpsons *as my framing device, since the show turned out to be the last big cultural tent of the pre-internet, prime-time-broadcast-TV era in mass media. It's hard to imagine another pop icon will ever speak as universally for its time as* The Simpsons *did in the 1990s, before the internet exploded the pop universe into a million tiny clusters of mutually exclusive interest.*

I wish it was the Sixties
I wish I could be happy
I wish I wish I wish that something would happen
                                        –"The Bends," Radiohead, 1995

ON THURSDAY, JANUARY 21, 1993, around 8:20 P.M. EST, I was standing on the edge of the dance floor at a campus pub called Alfie's in Kingston, Ontario, with a glass of cheap draft in my hand. Every seat in the joint was taken, the chairs and tables all jumbled into a kind of auditorium arrangement, and all eyes were fixed on the big-screen TV set up on the dance floor itself. The third and final act of Episode 9F11 of *The Simpsons* ("Selma's Choice") was just beginning, and there was a certain expectant tension in the room.

Now, 9F11 had already had some crowd-pleasing moments. The premise of the episode was that one of Marge's aunts, Gladys, had died a bitter spinster, setting a panicked Selma (one of Marge's ghoulish twin sisters) on a quest to have a child before her biological clock runs out. The episode had opened with a TV commercial for Duff Gardens—a theme park inspired by Springfield's favourite brew—that showed the Duff "Beer-quarium," an enormous mug of beer full of "the happiest fish in the world." (This joke played especially well with the Alfie's crowd, with hooting and cheering accompanying the image of one fish, cross-eyed and smiling, bumping repeatedly into the glass.) Later, as Selma set about the doomed task of finding a father for her child—via video personals, random passes at assorted minor characters, and a visit to the sperm bank—9F11 had filled in with the usual grab bag of great gags. Selma showed her sexy side by tying a lit cigarette in a knot using only her mouth; the Sweat hog whose sperm was available for purchase turned out, to everyone's disappointment, not to be Horshack; and, in a stellar example of *The Simpsons'* ability to condense note-perfect parody into a few short seconds, another TV ad for Duff Gardens featured a brief snippet of the teen variety act "Hooray for Everything!" singing a saccharine bastardization of Lou Reed's "Walk on the Wild Side," in a wonderfully silly send-up of Up With People. All in all, it'd been a solid episode through those first two segments, and certainly no one nursing their beers through the second commercial break that night could be completely disappointed.

By January of 1993, however, the crowds that gathered to watch *The Simpsons* around North America had come to expect not just *solid* but full-on *transcendent.* In Alfie's and a thousand other campus pubs, in dormitory common rooms and rec rooms and living rooms, in local taverns and sports bars (for *The Simpsons* was, by this time, the kind of TV event you wanted to share with your peers), the expectations ran high. It was midway through *The Simpsons'* fourth season—and about two years into what was, as I'll explain more thoroughly in a minute, a five-or-six-year run of pop-cultural virtuosity that was by far the most important cultural institution of the last decade—and *The Simpsons* had become something far more than a clever cartoon. In the fall of 1992 alone, at least every other of the eleven episodes aired—if not every episode aired—had been an instant classic. What's more, the episode that had aired the previous Thursday (9F10: "Marge vs. the Monorail") had been nothing short of a masterpiece.

All of which is to say that for many of us watching that Thursday night—at Alfie's and elsewhere—the critical bar had been set awfully high, and our collective anticipation was palpable. This new episode had had its moments, as I said, but it hadn't rocked the house to its very foundations. And we'd all learned that *The Simpsons* was capable of doing just that—consistently *and* thoroughly.

The Simpsons came back on, and the crowd at the pub went quiet. Because Homer was sick—he'd been picking away at a rotting ten-foot hoagie for weeks—it had fallen to Selma to take Bart and Lisa to

Duff Gardens. Chuckles from the crowd as Bart and Lisa point out four of the beer-bottle-costumed Seven Duffs: Tipsy, Queasy, Surly and Remorseful. A few quick cuts later, Bart, Lisa and Selma are poking around a souvenir stand. Bart approaches a display of clunky sunglasses. A sign reads: "BEER GOGGLES: See life through the eyes of a drunk." All at once, the pub shakes with a single great roaring laugh. It's as if a train is suddenly there in the room, its horn blaring, without any Doppler-effect build-up beforehand. It's like a force of nature, this laugh, spontaneous and open-mouthed and *enormous*. It nearly drowns out the next line: Bart puts on the beer goggles and turns to Selma, who has morphed fuzzily into a voluptuous babe, posing seductively. "You're charming the pants off of me," she says in a sultry voice. The laughter at Alfie's seems then to expand out exponentially, becoming a rollicking, almost deranged sound. People are doubled over, have tears streaming down their faces, are pounding tables with fists. It's as if that single gag was written for precisely this audience. It is beyond perfect, an act of clairvoyance, a sleight-of-hand feat in which some TV-writer wizard has invaded the brains of everyone in the bar, rooted around for just the right common reference, and then brought it flawlessly to life with a deft wave of a comedic wand. The last few minutes of the show play out to continued laughter.

Understand: By the standards of *The Simpsons'* Golden Age (spring 1991 to spring 1997, roughly), this was not an extraordinary episode. This was par for the course. This was the minimum payoff that could be expected *every week*. In subsequent weeks and years, *The Simpsons* would maintain a level of quality so high—would brew a mix of killer one-off gags and laser-guided social satire and robust character development and pure comedic joy into a potion so intoxicating—that it would become, for example, the primary metaphor I would use in conversations with most of my close friends and colleagues. And I know I'm not alone in this. There's a tendency, when attempting to discuss the broad strokes of a pop-cultural era, to lapse into a royal *we*—to invent a mass phenomenon out of one's own passion for the subject. I might well lapse into it on occasion hereafter, though I'm fighting it. But with respect to this particular subject, I'll embrace the *we*, because here it is accurate. Western culture has in recent years become an irredeemably fragmented thing, counted in webpage hits and record sales, endlessly quantified and analyzed and synthesized and then co-opted and corrupted by advertisers, focus groups, test audiences, pollsters, pundits, and on down the line, all the while changing so quickly and in so many directions that it has never really been nailed down. (Maybe no cultural moment ever really is.) But watching *The Simpsons,* all those scattered slivered *I*s became *we*s, if only for thirty minutes each week (more often after the show went into syndication). We were being defined by the show. Shaped by it. Even united by it, or as close to that state as we came, anyway. If there was a single cultural signpost broadcasting the emergence of a generation/era/movement/whatever, a monolith to a widespread yearning for progress, truth, honesty, integrity, joy, a final goddamn period at the end of every vacuous corporate press release and cloying commercial script and prevaricating political soundbite—it was *The Simpsons.*

To the uninitiated, *The Simpsons* might seem an unlikely vehicle for social change. It was on the crassest of corporate TV networks—the bottom-feeding Fox network—and its characters were shamelessly merchandised and frequently used as product spokespeople. But the show also gnawed relentlessly at every hand that ever fed it. (This was owing, incidentally, to a unique clause in the show's producers' contract that forbade any meddling whatsoever by the Fox suits.) It was never the most popular show on television, but its strong ratings among 18-to-34-year-olds for more than a decade (and, at the peak of its popularity, everyone under the age of 50) attested to the depth of its resonance with younger viewers and meant that, for example, it could charge the same advertising rates as a Top 10 show. And a recent wave of scholarly interest (cultural studies courses at a host of campuses in Canada, the U.S. and the U.K., thick tomes on the show's overarching philosophy and religious significance) adds to a legion of online enthusiasts who have created some of the longest-running and most exhaustive fantasies in cyberspace. I'm hoping, in short, that we can all agree that no mass cultural phenomenon since *The Simpsons* debuted in 1990 has had the impact that it has.

On the surface, the show's popularity derives from a source no different from that of any old comedy: It is extremely funny. *Funny,* though, is an extremely broad term, encompassing anything from slip-on-a-banana-peel sight gags, *Three Stooges* horseplay and Farrelly brothers gross-out jokes to the dark satire of, say, Stanley Kubrick's *Dr. Strangelove,* the political humour of Dennis Miller and P.J. O'Rourke, and the bizarro comedy-as-performance-art of Andy Kaufman and Tom Green (not to mention the comedic tradition in literature that dates back to Shakespeare).

The singular genius of *The Simpsons* is that it manages, at its best, to make use of the whole schmeer, sometimes even simultaneously. Every episode of *The Simpsons* is densely packed with straightforward gags: sight gags in the grand tradition of Warner Bros. cartoons, absurdist one-offs, parodies of pop-culture icons, self-referential and metatextual riffs for the postmodern crowd. Sometimes, single lines from *The Simpsons* stand on their own against anything Milton Berle or Woody Allen ever came up with. "In a way, you're both winners. But in another, more accurate way, Barney is the winner," a NASA scientist explains to Homer and Barney at the end of their astronaut training (Episode 1F13: "Deep Space Homer").

At other points, the myriad comedic instruments in the show's repertoire combine one on top of the other, building into a kind of symphony of one-upping humour. Anyone who has seen, for example, the sequence in 1F13: "Deep Space Homer" wherein a series of misfortunes befalls Homer and his space shuttle co-pilots, culminating in a scene of the space capsule turned into a bobbing mess of Ruffles chips, red ants and astronauts, accompanied by a James Taylor ballad, which scene convinces faithless TV anchorman Kent Brockman to hail the imminent arrival of our new leaders ("a master race of giant space ants")—yes, anyone who has seen that sequence knows just how elaborate, and how beer-spewing-out-your-nose funny, a series of gags of this sort can get.

The Simpsons would be a memorable show even if it did nothing more than make its fans laugh as often as it does. But it's the show's satirical subtext—its razor-sharp, surgically precise attacks on braindead, hypocritical power of every stripe—that truly distinguishes it from other comedy of its calibre. *Seinfeld,* for example, rarely managed to transcend its own goofily myopic little universe. *The Simpsons,* though, was a stealthy, subversive smart bomb sitting in the middle of primetime. Disguised as an inconsequential children's confection—a mere cartoon—*The Simpsons* would erupt each week in a series of spectacular satirical detonations, much to the increasing joy of a rabid fanbase that had always suspected the powers that be were a pack of cheats and liars, and now had proof.

The very bedrock of the show—its recurring characters, its setting, the basic premise of the series—is satirical. The show builds from the founding idea that the Simpson family are an average American household; early raves about the show frequently pointed out that the Simpsons seemed much more realistic than the flesh-and-blood people populating live-action sitcoms. This "realism" consisted of a world in which no authority was legitimate, no leader uncorrupted. The parents were all bumbling, the cops all corrupt. The celebrities—including guests from our world—were all grotesquely vain, the businesspeople veritably oozing pure evil. And yet it struck deep chords around the English-speaking world because—like all great satire—it rang true. We *did*—and do—live in a world of corrupt authorities, clueless leaders, rapacious businessmen. It wasn't just funny; it was true.

DURING THE SIX YEARS or so when *The Simpsons* was at its peak, it verged on infallibility, finding the last word on seemingly any subject it chose to seize upon. It follows from the phrasing of this assertion that I'm one of those fans who believes the show is past its prime. This is, to my mind, undeniable though not particularly tragic, and certainly not criminal, as some bitter former fans might assert. Far more remarkable is that the show—an enormous undertaking involving dozens of producers and writers (not to mention hundreds of animators)—managed to function at that high a level for that long. The Beatles, after all, were but four men, and childhood friends at that, and their peak creative period was six years long only by the most generous of estimates. Indeed, you'd be hard-pressed to find *any* collaborative creative enterprise *ever* that has had a run as long and as consistently ingenious as *The Simpsons* did.

A key element of the show's brilliance—and of its enduring popularity, particularly online—is its remarkable depth of detail. *The Simpsons* online community—so rabid and temperamental that it's been repeatedly parodied by the show itself—is an extraordinary achievement in itself. In addition to the expansive generalist sites, the show has spawned websites dedicated to every major character, websites built solely on bit characters, sites that obsess about the *Simpsons* world's geography and sites that track only the show's many references to Canada or Stanley Kubrick. At its peak, the community had a sort of hierarchy centred around the alt.tv.simpsons newsgroup and the Simpsons Archive website (snpp.com), both of which were run by a politburo of diehards who would meticulously evaluate and catalogue every episode, even deliberating on whether a particular line or shot constituted a reference or a mere coincidence. This sort of obsessiveness is now the web's stock-in-trade, but I'd hazard a guess that the *Simpsons* fansites were among its pioneers. And, more importantly, it was one of the only pop-culture icons that rewarded its fans with such a wealth of references and asides and in-jokes to chronicle. In fact, there's a sort of tradition—dubbed "Freeze-Frame Fun" in the Simpsons Archive episode guide—that evolved from the show's writers' knowledge that not a single detail they included would be missed by the devout.

Here is one of its most extreme examples, from Episode 2F06: "Homer Bad Man," in which Homer is accused of sexually assaulting his babysitter: A *Hard Copy*-esque current-affairs show, *Rock Bottom*, is forced to issue a formal apology to Homer for sensationalizing the accusations against him. *"Rock Bottom* would like to make the following corrections," the host says. A white blur of microscopic text then scrolls up the Simpson family's TV screen. If you taped the show and paused it at that moment afterward—as the show's hardcore fans inevitably did—you were treated to a lengthy and hysterically funny list of untruths.

Samples: "Styrofoam is not made from kittens"; "Our universities are not 'hotbeds' of anything"; "Ted Koppel is a robot"; "Bullets do not bounce off of fat guys."

Also: "The nerds on the internet are not geeks."

In Episode 2F10: "And Maggie Makes Three," Lisa asks her father a question about Maggie's birth, prompting Homer to reminisce about the event. "It was a tumultuous time for our nation," Homer recalls. "The clear beverage craze gave us all a reason to live. The information superhighway showed the average person what some nerd thinks about Star Trek. *And the domestication of the dog continued unabated."*

EVENTUALLY, THE STORY OF A few crazy, obsessive kids spinning their favourite hobby into a freewheeling, world-beating company with a multi-million-dollar IPO would become an archetype, and then a cliché, and finally—after the bubble burst—a cautionary tale. In 1995 and 1996, though, it was a singular thing, a brand-new bauble to stare at in wonder. It was simply marvellous. Just plain fun. Yahoo!

I mean, *dig* these guys: David Filo was the monk, the one who slept under his desk and kept driving his sputtering 1980 Datsun even after Yahoo!'s IPO in April 1996 made him a millionaire. Chih-Yuan "Jerry" Yang was the flamboyant one, the go-getter. He had the big house in Los Altos and the card that said his corporate title was "Chief Yahoo!"—what a gas!—and you just knew he'd be able to sell the *whole world* on this thing.

And the blazing seat-of-your-pants gall of it all, you know? What were these kids, twenty? Were they even old enough to drink yet? (Actually, Filo was 29 and Yang was 27 at the time of the IPO, but never mind.) Never mind at all, that's right, because this was *it,* man, this was the whole damn stake-your-claim, grab-the-brass-ring, go-west-young-man American dream here. This was the wagon train to the New Frontier, and happy days are here again! And there was such great *detail,* too, so many Horatio Alger touches. Stories about the rise of Yahoo! inevitably revelled in the suburban-kid mediocrity of it. Yang and Filo had been Stanford computer geeks working out of a trailer. They were always surrounded by a pile of fast-food containers. Later, when they got a real office, there were toys and computer games, as well. Late-night coding sessions, vats of caffeinated drinks, a foosball table, pizza boxes everywhere like floor tiles. And they were millionaires!

How'd they do it? That's the best part: They were slackers-made-good, is all. They liked playing around on their computers, but it was hard to keep track of where everything was as this World Wide Web thing kept growing. So they decided to try and organize it somehow. Turned out that was *exactly* what everyone needed as soon as they went online, because this web was just ridiculously vast, impossibly complex. It was a universe.

You wouldn't believe all the stuff that's out there.

For the few years leading up to the emergence of Yahoo! and the rest of the first wave of internet companies—and even, concomitantly, for a couple years after—the dreamers managed to get some headlines too. What's more, for a while there, no one even thought of them as dreamers. They had more experience with the internet than anyone in officialdom, after all, and so they were courted, consulted, invited to CNN talk shows.

Take John Perry Barlow. He was an unlikely poster boy for the digital revolution: an aging hippie, a one-time composer of lyrics for the Grateful Dead, a long-time Wyoming cattle rancher. Drawn to the internet even before the birth of the web by a group of Deadheads who were congregating online at the WELL—the Whole Earth 'Lectronic Link bulletin board that became an incubator for techno-libertarian dreamers—Barlow was, for a time in the early 1990s, a fervent evangelist for all things internet. He became a media darling, always ready to step up and repeat his assertion that the internet was not just the biggest thing since Gutenberg's printing press but "the most transforming technological event since the capture of fire." To Barlow—unlike many of the early internet libertines, who eventually picked the free market as their vehicle for revolution—the real power of the new medium was its ability to create community. Barlow saw the first online discussion groups as the potential inheritors of the ideals of the sixties counter-culture, each one a renewed society in embryo. All that was needed was to keep the forces of corporate greed and legislative fear out—to keep the internet free. "I felt I had found the new locale of human community," Barlow wrote in 1995.

Barlow was on to something, of course. The internet's countless communities—whether current-affairs sites like Slashdot or Britney Spears fan groups or cybersex chatrooms—have become some of the digital world's most beloved and enduring places. In a sense, the communal aspect of online communication that had so awed Barlow and his semi-utopian contemporaries was the flipside of the same coin that was the target of the Yahoo! boys' obsession. To wit: On the one hand, the internet was unfathomably vast, encompassing enormous stretches of territory—and not just physical space, but limitless tracts of cultural and psychological terrain, as well. On the other, the new medium was proving to be as deep as it was wide. Stop in any one place, and you'd soon realize that that spot—whether a web ring cataloguing *Simpsons* references or a discussion group bent on changing the world—contained a microscopic universe as complex as the whole.

It was all too much. It could induce a woozy, fearful condition sort of like vertigo, especially in those raised on analog. You'd find yourself so discombobulated you'd have to get your kid to explain what was happening.

ONE OF THE INTERNET AGE's most widely held myths has to be the belief that young people are fundamentally more knowledgeable than older people about digital technology, possibly even possessing some innate computing skill that anyone who reached adulthood before around 1980 could never possibly develop. There might well be some small kernel of truth to this, but the real secret is simply that those who have grown up using computers are not *afraid* of them. Because of this relative comfort, young folks were given greater control of the internet than any other medium going, and because of *this,* the culture of the internet developed a certain youthfulness.

This points toward one of the fatal miscues of the overwhelming majority of the doomed dotcoms: their belief that because people were doing certain things online, they wanted to do *everything* online. It wasn't, that is, that young people wanted their whole culture delivered to them online; it was that there were certain kinds of culture that they could find only on the internet. And this, moreover, was partially because the rest of the culture was so thoroughly dominated by a hyper-corporate, top-down approach that many young people felt completely alienated from it. It had very little to do with them. But the internet—which, because of their "innate" skill, they'd stuffed full of the stuff they *could* relate to—was seemingly all theirs. The internet was not, as the latecomers to the dotcom feeding frenzy seemed to believe, the *actual* thing that all these people were yearning for; it was just the metaphor.

This was what initially made the net so appealing: It wasn't on CBS or Global. It wasn't brought to us by Pepsi or Ford. It wasn't controlled by the regulatory arm of any particular government. And it was emphatically *not* better when the Boomers wore their hair long.

In Episode 3F21: "Homerpalooza," Homer takes Lisa and Bart to the Hullabalooza music festival in a desperate attempt to convince them he's cool. They arrive at the entrance to the festival grounds—which is plastered with ads—and Lisa is excited. "Wow!" she says. "It's like Woodstock, only with advertisements everywhere and tons of security guards."

THE ATOMIC BOMB ON NIRVANA'S *Nevermind*—the song that would almost single-handedly obliterate the hair-metallic, candy-popping, brain-dead eighties aesthetic and usher in several years of angry rock sludge—was, of course, the opening track. "Smells Like Teen Spirit" opens with a melodic, hiccupping jangle of clean classic riff-rock. It bounces along like that for a full bar... starts to repeat itself... and then, just when you've started tapping your foot along to this pretty little confection... right at the midpoint of its second run... the bomb drops: four thundering explosions from the drums, three dense beats each, clearing the way for a ferocious roar from the rest of the band. And incinerating nearly two decades of middle-of-the-road Top 40 pap for good measure.

The music that would fill the bomb crater left by "Teen Spirit"— which would reluctantly adopt the moniker "grunge"—was unwaveringly gloomy, lending credence to the notion that the youth of the early 1990s were terminally pissed off, disaffected beyond salvation, cynical to the core. "Oh, well, whatever, never mind," Kurt Cobain mumbles at the end of the last verse of "Teen Spirit," seeming to confirm the rumours. Had a little more attention been paid to *Nevermind*'s seventh track, though, perhaps the moronic mainstream that Cobain so loathed would have noticed the yearning idealism at the heart of the music.

That song—"Territorial Pissings"—begins with a savage, tuneless reading of the chorus of the Youngbloods' feel-good hippie anthem "Get Together," sung by Nirvana bassist Krist Novoselic. (Novoselic, it warrants mention, would become a committed political activist after Nirvana's premature demise.) The sound of the vocal is fuzzy and distant, as if recorded onto a cheap tape long ago and long since forgotten. "C'mon people now, smile on your brother," Novoselic yelps. "Everybody get to*gether*"—his voice seeming to bottom out on this syllable, exhausted from the strain of making the sentiment sound plausible—"try to love one another right *now.*" Even before he's done, a single ominous note of feedback-drenched guitar invades. It repeats four times, waiting. The lyric ends, and a jagged wall of fuzz takes over, resolving into a lightning-quick riff. A double-time drum roll comes up underneath it, and then the song hurtles into blistering 4/4 punk. Cobain snarls his lyrics, building into a full-blown yell at the chorus: "Gotta find a way, a better way, I'd better wait…" By the end of the song, two minutes later, the yell has unravelled into a soul-sick howl. "Gotta find a way, a better way, I'd better wait…" Finally—pushed beyond reason, rendered subliterate by his rage—Cobain only screams. It is a brutal, desperate sound, clinging to hope by a thin rope frayed beyond salvation. A little over two years later, he would be dead by his own hand—horrific proof that the rope had finally snapped.

Gotta find a way, a better way, I'd better wait…

HERE'S A FUNNY THING ABOUT "Get Together" by the Youngbloods. It would be resuscitated at least twice more as the decade wore on: first as part of the soundtrack for the 1994 film *Forrest Gump* (a film whose central message was that the path to success in America was paved by blind obedience); and then, the following year, as the backing track for a Pepsi commercial. Of course, by the time of the Pepsi ad, only the most self-deluded and hyper-nostalgic of Baby Boomers could possibly have believed that there was a single shred of the sixties counter-culture's ideals—the ones being celebrated by the song—left to sell out. Not that it wasn't an eventful yard sale, mind you; in fact, the sort of people who dug Novoselic's butchering of one of its anthems had grown up smothered in the broad shadow that the sixties sellout had cast.

The first unmistakeable, billboard-sized sign that the sale was on had to be the 1987 Nike commercial that used the Beatles' "Revolution" as its soundtrack. Way back then, such a blatant corruption of a counter-cultural icon still had the power to piss people off. Boomer-aged pundits sounded off in every other newspaper about the heresy, and the Beatles themselves sued Nike, its ad agency and the record companies involved. Eventually, Nike agreed to stop airing the ad, but by that time, the sale-cum-revolution was on, and Nike was front and centre in its vanguard. For Nike *had* instigated a revolution of a sort: The "Revolution" ad was the first major action in its ultimately successful campaign to convince the world of the innate value of corporate brands. In the years that followed, Nike transformed itself from shoe company to pure marketing machine—a new-age business with few fixed assets and no manufacturing facilities whose primary value resided in a swoosh, a slogan and an attitude. In the years that followed, the very idea of nonconformity became little more than a marketing strategy (see, for example, Apple's "Think Different" ads), and the recasting of sixties rebellion as nineties consumption eventually ceased to ruffle many feathers. By the mid-1990s, both the Bank of Montreal and multinational accounting firm Coopers & Lybrand used Bob Dylan's "The Times They Are A-Changin'" in their ads, causing only feeble and ultimately impotent criticism. The bank ad, in particular—which depicted a choir of angelic children on a mountainside, singing about changing the system—may well have caused a great many *Simpsons* and Nirvana fans to wretch (or at least change the channel). But the idea that *any* icon, no matter how diametrically opposed to the corporate-consumer ethos it might appear, was not for sale—this notion had died years earlier, one of the first casualties in the glorious Nike revolution.

In Episode 5FO5: "Lisa the Skeptic," Lisa and her classmates go on an archaeological dig to a spot that is to become a parking lot for a new mega-mall. They uncover a winged skeleton that the Springfield populace embraces as a sacred idol. The angel disappears mysteriously, then reappears to lead the entire town to a hill next to the new mall, where it comes to rest beneath a sign reading, "The end will come at sundown." At sundown, it's revealed to be an elaborate publicity stunt: "the end of high prices" at Heavenly Hills Mall. Lisa berates the mall's developer: "You exploited people's deepest beliefs just to anoint your cheesy wares. Well, we are outraged, aren't we?" A distracted Police Chief Wiggum responds, "Oh yes, we're outraged. Very much so... but look at all the stores. A Pottery Barn!" And the townsfolk rush off eagerly into the mall.

A FEW STATS, FOR AN ERA obsessed with raw data:

Since 1987, there have been more shopping malls in the United States than high schools. An estimated 70 percent of Americans visit a mall at least once a week. The 1990s played host—famously—to the greatest economic expansion in U.S. history. Those years also ushered in that nation's greatest economic polarization at least since the 1930s: By 1999, average CEO compensation had ballooned to 475 times the average worker's salary, up from a factor of 85 in 1990 (by comparison, the average British CEO made 24 times the average worker in 1999). As of 1997, 40.1 percent of America's wealth sat in the pockets of its richest one percent, up from 20.5 percent in 1979. The fastest-growing city in America during the last decade was Las Vegas, and Nevada was the fastest-growing state; at the same time, Nevada had the largest number of residents living permanently in mobile homes. Manpower Inc., a temporary-employment broker, became America's largest employer in the 1990s. As of 2000, the U.S. ranked 139th among the world's 163 democracies in voter turnout. (In the May 2000 school board and criminal district elections in Fort Worth, Texas, fewer than five percent of eligible voters cast votes.) Kip Kinkel, the fifteen-year-old boy who killed two of his classmates and wounded 22 others in Springfield, Oregon, was—like between 1.5 and 4 million American kids—taking Ritalin. He was also—like about 800,000 other American kids—taking an anti-depressant. Ritalin use among Canadian children increased by 600 percent from 1990 to 1998. Overall, anti-depressant prescriptions (for adults and kids) in Canada increased by 63 percent from 1996 to 2000, the period during which the country participated most explosively in the economic boom that had begun a few years earlier in the U.S.

The overriding ideology you can sketch by connecting these dots goes by a variety of monikers: *late capitalism, consumerism, corporatism, corporate consumerism, corporate capitalism, free-market capitalism, liberal democracy.* It is sometimes summarized by the expansive umbrella term *modernity,* which seems the most accurate, though it opens up opponents of any particular point in the program to being glibly dismissed as flat-earthers. At any rate, it is this ideology—and the indifference verging on contempt that it shows toward any human need that cannot be bought or sold, that cannot be measured by polling data, that does not in some way contribute to per-capita GDP or the balance sheet of a multinational corporation—it is *this* ideology that created the preconditions for the rage of a Kurt Cobain, the angry sarcasm of *The Simpsons.*

These are the hairline fissures on the sleek surface of modernity.

In Episode 2F19: "The PTA Disbands," the teachers at Springfield Elementary School go on strike. The Simpson family discusses the imbroglio at dinner. Homer is angry that he's stuck taking care of his kids; Lisa rises to the teachers' defence. "Lisa," Homer explains, "if you don't like your job, you don't strike. You just go in every day and do it really half-assed. That's the American way."

THERE WERE FEW UNIFIED responses to the hollow prosperity of the 1990s, no tightly woven web of icons and events and symbols that could be condensed into the kind of tidy montage that tends to pop up, for example, in films about the sixties counter-culture. Many—perhaps most—merely bought in on some level. They got jobs at dotcoms (or in "bricks-and-mortar" businesses), collected stock options, bought houses in the suburbs, shopped in gargantuan "big box" stores, drank Starbucks coffee religiously. There were also widespread tendencies to either: a) disappear entirely into a cozy, sequestered corner of the culture, attempting to build a whole society out of whatever happened to be lying around there (viz. conspiracy theorists, Trekkies and other Trekkie-like subcultures, hackers, Phish-heads, attendees of drum-and-bass and *only* drum-and-bass parties, proprietors of internet fansites, etc.); or b) watch the whole parade from a comfortable ironic distance.

One of the undisputed kings of this latter stance is the satirical newspaper *The Onion*, best known for its online version. The site's deadpan parodies, whose ostensible goal was to relentlessly mock the very idea of "news," somehow managed also to become remarkably adept—certainly far more adept than any newspaper that reported on actual events—at cataloguing the variant strains of unease and dissatisfaction floating freely through the culture. As with *The Simpsons, The Onion's* approach—and its success—was predicated on the notion that the average consumer was deeply cynical about potentially anything that was happening in the world, from presidential politics to international conflict. But *The Onion* also spoke, frequently, of the numbness at the core of modernity. The headlines alone tell the story: "Aging Gen-Xer Doesn't Find Bad Movies Funny Anymore." "Twelve Customers Gunned Down in Convenience Store Clerk's Imagination." "Klingon Speakers Now Outnumber Navajo Speakers." "Fast-Food Purchase Seething with Unspoken Class Conflict." "Everything in Entire World Now Collectible." "Account Manager Fondly Remembers Day in College When Everyone Hung Out on Roof." "U.S. Dept. of Retro Warns: 'We May Be Running Out of Past.'" "New Crispy Snack Cracker to Ease Crushing Pain of Modern Life."

In the writings of the Situationists—the radical French philoso-
phers whose ideas were a primary force behind the Paris student up-
rising of 1968, and who are the acknowledged godfathers of the contemporary "culture-jamming" movement championed by *Adbusters*
magazine—modern capitalist society is described as a world of pure
spectacle, incapable of providing any real benefit or meaning to anyone outside the ruling bloc. Alongside calls to revolution, the situationists advocated political activities which would draw people's
attention to the fact of this spectacle, the absurdity of the society.
This is what *The Onion* does on its best days.

At all times, though, *The Onion* is also a shining example of a much
broader phenomenon in recent pop-cultural history. This is the one
somewhat vaguely—and, of late, dismissively—gathered under the banner of irony. But irony refers only to a subset of the bigger picture, that
wide array of attitudes, styles, modes of discourse and expression that
have emerged collectively as testament to the extraordinary sophistica-
tion of the brand of consumerism practised by (predominantly) young
people in the post-industrial West. Think here of the logo-parodying
T-shirt ("Enjoy Coca-Cola" becomes "Enjoy Cocaine"); of advertising
meticulously crafted to anticipate and derail cynicism about advertising (Sprite's "Obey Your Thirst" campaign comes to mind); of getting
your news via late-night-TV monologues; of the so-bad-it's-good movie; of the dominant tone of the discussions on metanews websites like
Slashdot and Plastic (cultural critic Thomas Frank has described this
"wiseass tone," perhaps a little too wiseassedly, as "the internet's greatest gift to civilization"). Recall that the Spice Girls movie, *Spice World*,
was knowing and self-referential and self-parodying. So are most beer
commercials. All of this, taken together, is what it looks like when a
culture moves from merely nibbling at its own tail to devouring it in
exquisite eight-course meals. The Renaissance Man may be dead, the
classical education all but forgotten, but our streets teem with professional consumers of the highest calibre. We know how our products are
packaged, how our media content is produced, and we will only buy it
if it lets us know that it knows that we know.

Perhaps this is why the dominant medium of the last decade—
feature film—rarely had much social or political impact outside the

entertainment industry itself. Mainstream movies became brilliant marketing and money-making vehicles, perfecting the art of the six-month promotional campaign and the product tie-in, ever more adept in the ways of selling themselves across demographic and geographic borders. But how many of them changed your understanding of the world at large? Meanwhile, the most resonant films were at least as interested in teaching us about film-making as they were in enlightening us about the human condition. Films like *Pulp Fiction, Swingers, Trainspotting* and *Being John Malkovich* may have struck deep chords and won rabid fans, but their reverberations were discernable only in brief trends in music, fashion and slang—and in the halls of the world's film schools. The Coen brothers made a series of films that masterfully reworked Hollywood's favourite myths. Wes Anderson's brilliant fables are set in their own universes. Even Steven Soderbergh's overtly political double bill, *Erin Brockovich* and *Traffic*, simply hammered home long-standing arguments ("large corporations are careless and cynical" and "the War on Drugs is a misguided failure," respectively), speaking most loudly and originally on the subject of how to turn TV-movie-of-the-week fare into stylish, artful cinema. There were good movies about making movies and bad movies about making money, but there were few films that spoke directly to their age in the tradition of *Easy Rider* or *The Graduate* or *Apocalypse Now* or *Heathers*—or even *Top Gun*, which, lest we forget, precipitated a dramatic rise in U.S. Navy enrollment figures in the mid-1980s.

There was one exception, a micro-genre I'll refer to for the sake of brevity (I know, I know: Why start now?) as the entropy movie. The genre hit its commercial apex with Sam Mendes' *American Beauty*, but it also includes Ang Lee's *The Ice Storm*, Robert Altman's *Short Cuts*, David Fincher's *Fight Club*, P.T. Anderson's *Boogie Nights* and *Magnolia*, several Richard Linklater films *(Slacker, Dazed and Confused* and especially *SubUrbia)*, and the collected works of both Neil LaBute and Todd Solondz. These films were often, but not always, set in faceless suburbs; what unites them, however, is not the physical but the *psychological* setting. Or, more precisely, the *moral* setting, for these are movies that exist in places all too familiar to us, places where the moral compass that should be guiding human interaction is either broken or warped or else entirely absent. In an entropy movie, characters have to

145

beat the shit out of each other or swap spouses at key parties in order to feel the most basic sense of community; they find belonging not in their homes but on the sets of porn movies; they engage in brutal emotional warfare with each other simply because they can get away with it. They live, as we do, in a society that's in the midst of a profound crisis of moral authority. This is the nugget of truth at the centre of all those ludicrous family-values uproars of the past twenty years: We have, as a society, forgotten what it is we believe in. Our priorities have become so arbitrary—and thus so skewed—that we have become like molecules firing off in all directions at the edge of a grand order as it collapses into disorder. Into entropy. We know when we are wronged, perhaps, but pay little mind to the idea of what's right. Which is the greater outrage in *American Beauty*—that the protagonist lusts after his daughter's friend? That he acts on that lust? Or that he, his wife, his daughter, his neighbours—all of them—have built worlds for themselves in which they can barely remember what it even means to be passionate? All, that is, except the strange, intense, self-confident Ricky with his digital camera. It is Ricky who sets *American Beauty* apart from most entropy movies, because it is he who points the way out. *Fight Club* ends with nihilism, *Boogie Nights* with chaos, *The Ice Storm* with quiet desperation. But in Ricky's simple video of a plastic bag dancing in the wind—and such a pointedly ugly and banal object, too, such a powerful symbol of synthetic, antiseptic, generic modernity, a plastic bag—in his video, there is a bold, simple declaration: This is beautiful. Even this. If you look the right way, anything is. Care about this, or about anything else, but *care,* at least. Start with that.

A FACTOID: FOR THREE DAYS in August 1998, Lemonwheel—the concert/festival/universe that marked the end of Phish's summer tour—was the second-largest city in the state of Maine.

THIS WAS ANOTHER WAY to respond: Build a parallel universe.

The name Thomas More gave to his ideal society—*Utopia*—was an intentional pun. It meant both "good place" and "no place." Though the term *utopia* has long since lost this ambiguity, Lemonwheel—and a great many other festivals and events and parties in the past decade—seemed implicitly to understand (and embody) the double meaning.

Lemonwheel was an extraordinary event, a mass gathering of about 65,000 people from wildly diverse subcultural tribes (among them aging hippies, neo-hippies, bikers, computer geeks, skate punks, frat boys, rave kids and at least one orthodox-Jewish klezmer band) on an abandoned air force base a full day's drive from any major urban centre. Besides the ostensible main attraction of the event—the music of the jam-rock band Phish, delivered in six sets over two days—the festival included dozens of enormous art installations, a Zen rock garden, countless impromptu raves and mini-concerts, and a commercial district of temporary food and merchandise stalls that ran the length of both runways. It was a self-contained world, with its own customs and language (including a staggering array of slang terms for marijuana), its own values and rules. With the exception of the unmarked black helicopters (almost certainly DEA) that circled the site throughout the daylight hours, Lemonwheel was entirely self-governing. It was a utopia. It was, for all in attendance, a good place. And it was also no place: It was understood by the band, its fans and the state troopers at the gate that it would be gone after three days, that it would be dismantled, that everyone involved would return to whatever world they actually lived in.

The same structural and temporal limitations governed nearly all of the most exciting socio-cultural movements of the 1990s. The Burning Man festival came but once a year, for a specified time period, and one of its core rules stated that it leave absolutely no trace of itself behind. Rave scenes, which sprung up in nearly every urban centre in the Western world at some point in the last decade, were limited almost exclusively to weekend parties in constantly changing locations. Online communities rarely even attempted to leave the noplace of cyberspace. And even the decade's most important political movement—anti-Globalization—has thus far been unable to sustain itself outside the parameters of concise events organized by its opponents (not to mention its chronic difficulty even in maintaining any kind of consensus for the duration of those brief summits). Nonetheless, these events provided potent doses (even overdoses) of essential human needs—political involvement, spiritual fulfillment, social freedom, the transcendent joy of communal unity—that couldn't be found in adequate measure in day-to-day society.

These were desperate utopias.

ALLOW ME HERE AN ASSERTION that would be impossible to verify scientifically: The point of the pop-music compass upon which the largest number of these utopian vectors converged was the 1997 album *OK Computer* by Radiohead. This is not to say that it was the album that the largest combined total of ravers and Phish-heads and anti-corporate crusaders and the rest held dear (though that might also be true); rather, that it was the mainstream pop album that best encapsulated—more or less as it was all happening—the disaffected yearning of all of these atomized tribes.

OK Computer opens with a sort of rebirth: the near-death experience of a car crash, with salvation courtesy of technological progress ("Airbag"). Over the loosely linked eleven tracks that follow, it's as if the accident has caused a subtle but potentially fatal case of schizophrenia, as the songs' narrators lurch desperately from scenes of madness and terror and creeping dread to visions of escape, retribution and narcotized peace. Throughout, the marvels of the modern world that saved lives in the opening track (neon signs, intastella bursts, the airbags in fast German cars) are recast as cogs in a soulless, inescapable machine. Lightning-fast transportation only leaves us let down and hangin' around; we are haunted by buzzing fridges and detuned radios; alarms are not only incapable of ensuring our safety but are themselves a clanging tyranny. Vast swaths of *OK Computer* sound like the kind of twitching, nerve-shredding blowout that must precede road rage, or a killing spree, or the nervous breakdown of a mid-level executive who winds up dead in a suburban garage full of carbon monoxide. As with Nirvana's misconstrued *Nevermind, OK Computer* was so awash in anguish and despair that it was easy to miss the torrid, cathartic rage at its core. The latter was lifted explicitly to the surface by the brilliant video for "Karma Police." The video tells a muddled tale: Thom Yorke sits in the backseat of a car that rolls slowly down a dark country road. A desperate man in corporate-issue white dress shirt and slacks is trapped in the car's headlights, arms flailing as he runs, his doom seemingly assured. "This is what you get," Yorke sings, his voice barely above a whisper, but firm and self-assured. "This is what you get when you mess with us." The man in the headlights stops. The car stops and reverses, leaving a trail of leaking gas between it and the man. The man sets it alight, and the car is engulfed in flames. The backseat is now empty. Only the driver of the car—from whose perspective the entire video is seen—is destroyed. The viewer is left with the certainty that righteous vengeance was carried out. But upon whose guilty soul? Who was driving the car?

Amazingly, *OK Computer*—this seething mass of musically adventurous, radio-unfriendly nightmares and twisted revenge fantasies—became immensely popular, as well as one of the most widely acclaimed albums of the last decade. It was one of those albums that seemed to speak directly to its time, giving voice to all the fears and doubts its audience had as they flew through their lives at internet speed.

In Episode 5F09: "Trash of the Titans," Homer is elected sanitation commissioner of Springfield, and promptly spends his entire annual budget in the first month. To raise new funds, he turns an abandoned mine on the outskirts of the town into a giant garbage dump, accepting trash from cities across the country. In short order, Springfield is overrun with garbage—it spills out of sewers, erupts out of the ground, pours out of fire hydrants. Eventually, the entire town is moved "five miles down the road" to "solve" the problem.

IN OCTOBER 2000, TORONTO city councillor Olivia Chow screened Episode 5F09 of *The Simpsons* in council chambers. The audience consisted of the entire Toronto City Council, plus a packed public gallery. The occasion was a marathon debate about the city's plan to "solve" Toronto's garbage problem by trucking its waste to an abandoned mine near Kirkland Lake, 600 kilometres north of the city. It was a thing of rare, incomparable beauty—an inspiring moment in North American political life in the fall of 2000. At the time of Chow's broadcast, Canada was soon to embark on an unwanted, irrelevant federal election that would eventually see the lowest voter turnout in the country's history (a dismal 61 percent)—and result in the re-coronation of a federal government with no real vision for the country that had come to power merely because its predecessor was so widely loathed, and had remained in power because no credible nationwide alternative had emerged. The United States, not to be outdone, boasted an even lower turnout (51 percent) in its presidential election, which would famously result in a virtual tie between two candidates who inspired fierce debate between American voters only on the issue of which one seemed less inspiring. And what did any of it matter, anyway? In the fall of 2000, the economies of both countries (as well as most of the West) were on high-speed cruise control.

Do you remember just how smug and complacent and frivolous was the summer and fall of 2000? Do you recall, for example, that the "whassup guys"—that is, a couple of the actors who became famous for screeching "Whassup?" at each other in a beer commercial—went on tour? Of course, it was easy to miss the whassup guys' tour, what with the enormous excitement engendered by the first season of *Survivor,* which was front-page, lead-story news for most of August. Earlier in the summer, a rerun of the drama *Third Watch* earned higher ratings in the U.S. than the acceptance speech of Republican vice-presidential candidate Dick Cheney.

Again: What did any of it matter, anyway? As right-thinking people knew, history ended in 1989 with the publication of Francis Fukuyama's essay "The End of History?" in the journal *The National Interest.* Fukuyama's argument was simple enough: The collapse of communism represented the death of the last remaining alternative to "Western liberalism." All that was left was to get fat and tidy the lawn. And what happened in the 1990s to prove him wrong? The great stories of the decade were the ones about consolidating Western liberalism's victory and expanding its scope. There's a McDonald's *where* now? Beijing? Not only can you watch CNN in Timbuktu, dude, you can send an email from there to the South Pole. Then maybe use e-Trade to buy some stock in a Tuvalu-based dotcom run by an ex-pat Australian Buddhist, a Taiwanese math whiz and a game-theory expert from Kazakhstan. "For our purposes," wrote Fukuyama, "it matters very little what strange thoughts occur to people in Albania or Burkina Faso, for we are interested in what one could in some sense call the common ideological heritage of mankind." The war's over. The good guys won. Pass the eXtreme Doritos.

As long as you were lucky enough not to be born in the southern hemisphere or the rougher parts of Houston, then they *were* days without history. So much was happening, so much change, but nearly nothing we could hold onto, few signposts: just a mad rushing flux. We stood back from it, detached ourselves, and watched the parade.

So Fukuyama *was* right, but not in the way he thought he was. It's true that it doesn't matter what's being said in Burkina Faso or

Albania; what matters (and what Fukuyama and the free-market triumphalists failed to recognize) is that Burkina Faso and Albania—and Tuvalu and Timbuktu and Afghanistan—are part of a single multi-faceted entity created by global integration. They are part of the conversation, whether we (or they) like it or not, and if we're too busy shouting congratulations at each other to listen to what they're saying, they may just force their way in.

Put another way, history was storing its energy. It was building up for something massive. It abhors vacuums. And then it came, flew in out of a sky so clear and blue that it was a key part of the story, a detail that everyone felt compelled to note.

It was such a beautiful day. Not a cloud in the sky.

In Episode 2F11: "Bart's Comet," Springfield is under imminent threat of total destruction by a comet hurtling toward the earth. After the only bridge out of town is destroyed, the U.S. Congress springs into action, tabling a bill to evacuate Springfield. Just before the vote, a Congressman adds a rider to the bill, aiming to provide $30 million for "the perverted arts." The bill is soundly defeated. Cut to news anchor Kent Brockman, who says to his viewing audience: "I've said it before and I'll say it again: Democracy simply doesn't work."

WITHIN A FEW SHORT DAYS OF the World Trade Center collapse, America's leaders had carefully surveyed the damage. Had considered its enormous impact on everyday life. And then had risen as one in a brave chorus, imparting their sage advice to a shaken nation. The advice? Go shopping. By the tragedy's two-week anniversary, President Bush had called for "continued participation and confidence in the American economy." New York Mayor Rudolph Giuliani advised, "Just spend a little money." Vice-President Cheney was urging a "normal level of economic activity." Senator Bob Graham explained that buying a new car was "an act of patriotism." The president's brother, Jeb, added shopping, going to a restaurant and taking a cruise to the list of patriotic acts. Senator John McCain, beloved for his straight-talking ways, cut succinctly through the politicking. "We need," he explained, "to spend money." Elsewhere in a shaken Free World, little was said beyond heartfelt but ultimately hollow platitudes about standing "shoulder to shoulder" with our American comrades.

We were all finally ready to move beyond the trivialities of *Survivor* and day-trading and a new pair of Nikes. But in the weeks and then months after September 11, not a single one of the great leaders of our peerless West—these lands of the free, cradles of civilization, beacons of liberty and justice, wellsprings of unparalleled opportunity and prosperity—not a single one could articulate even a vague notion of what it was we were defending. All agreed it had to be defended, but none had a clue what would help on the homefront. Except this: We were being told—in emphatic terms—that the thing we were defending and the way to defend it were one and the same; we were the great globalized Republic of Buying Stuff, and so we should bravely go on buying stuff to protect our right—our most basic, most cherished right—to buy more stuff.

We will never know just how great an opportunity was lost, how much passionate momentum squandered. Here were the people of virtually the entire world—and certainly the entire First World—rising as one, ready to sacrifice, wanting to help. Who knows how many oversights could have been corrected, inequalities eliminated, hypocrisies inverted? Who knows what glorious civilization could have emerged from the ashes of those towers? One thing is for certain: Our leaders didn't, and don't. They have failed us completely in this regard—particularly those of the United States of America, that self-proclaimed last best hope for humanity. America's primacy of place and supremacy of power in world affairs, be they economic, political, military or cultural, has never been more apparent than in the wake of September 11, when, as if in some half-witted Hollywood movie, the whole world looked to America for direction. And was told to go shopping.

It was a response so repulsive, so utterly empty of insight and so overflowing with self-interest, that it shook my faith in Fukuyama's beloved "Western liberalism" to its very core.

AT 10:13 A.M. PACIFIC TIME on September 11, a post was made to MetaFilter, an online current events weblog. It read:

Things fall apart; the centre cannot hold;
Mere anarchy is loosed upon the world,
The blood-dimmed tide is loosed, and everywhere

The ceremony of innocence is drowned;
The best lack all conviction, while the worst
Are full of passionate intensity.

This is an excerpt from "The Second Coming" by W.B. Yeats. It is a succinct, accurate description of the people who organized and executed the attacks of that day. It seems to me just as accurate a description of the state of affairs in the West throughout the past decade. The inspirational forces in our culture—the internet, utopian festivals, extra-political movements aimed at curtailing corporate power, environmental devastation, marketing excess—were a million tiny atoms left isolated, swirling randomly in an agitated whirlpool as the centre lost its grip. Meanwhile, stock-market speculation and dotcom money-grabbing and corporate power-mongering and bloated mega-spectacles (on film, on compact disc and in print) illustrated the passionate intensity of a morally and creatively bankrupt mainstream.

I frequently came across another high-minded quote in the days after September 11: "An eye for an eye only makes the whole world blind." This one is from Mohandas K. Gandhi, and was often cited by opponents of the war on terrorism's excesses. It was, to my mind, the wrong Gandhi quote. Try this one instead: "You must be the change you wish to see in the world."

I might be wrong
I might be wrong
I could have sworn I saw a light coming on

I used to think
I used to think
There is no future left at all
I used to think

Open up, begin again
—*"I Might Be Wrong," Radiohead, 2001*

PICK A STARTING POINT. Maybe a story appears in your newspaper, a dull story about some arcane legislation. I'll begin there.

In 1995, the Organisation for Economic Co-operation and Development (OECD) initiated talks aimed at drafting an agreement that would normalize investment practices for all of its twenty-nine member nations—which nations represent more or less the entirety of the First World. By early 1997, there existed a draft of the pact, and it had a name: the Multilateral Agreement on Investment, or MAI. Some members of the Canadian delegation, however, felt that the MAI was scarily over-ambitious. In a momentary triumphalist lapse, Renato Ruggiero, then head of the World Trade Organization, described the document thusly: "We are writing the constitution of a single global economy." So the Canadians leaked it to the Council of Canadians, a nationalist public-interest group, who passed it on to another, larger public-interest group called Public Citizen (an organization founded, not incidentally, by Ralph Nader). Public Citizen posted the document verbatim on its website. In a few short months, a coalition of more than 600 non-governmental organizations around the world had united to oppose the MAI. The document—and the growing opposition to it—had made it into the headlines of the mainstream press by early 1998. OECD talks on the agreement were pushed back in April 1998 to give the organization's members time to consider its ramifications in more detail. Negotiations were abandoned altogether that December. The MAI was dead. Some commentators in the business press insisted that in-fighting between the OECD's members was what killed it, but most—including *The Economist*—admitted that the protest movement that had begun on the internet was the primary architect of its demise. The loose coalition that had formed included a broad spectrum of left-leaning groups, from major labour unions to environmental groups to human-rights watchdogs. It would form the nucleus of a broader protest movement that shut down a major WTO meeting in Seattle in late 1999 and transformed a series of subsequent meetings by global capitalism's top proponents from orderly closed-door affairs into noisy street theatre.

As the MAI began its death throes—but before the Seattle protests—Stephen J. Kobrin, a top official from the Wharton School at the University of Pennsylvania (one of America's top business schools), took to the pages of the influential journal *Foreign Policy* to ask, incredulously, "What is going on here?" Why had this "arcane treaty" given rise to such "broad, intense, and vehement opposition"?

Kobrin then went on to examine the details of the treaty in language as arcane as that of the MAI draft itself. He needn't have bothered. The facts of the MAI were largely incidental to the breadth and vehemence of its opposition.

That opposition emerged from a mood shared by countless thousands, a sense that something wasn't quite... right... with the state of our society:

- those aging hippies who actually *had* been bothered by the commodification of their most cherished icons
- the ones who watched *The Simpsons* as much for the social commentary as for the yuks
- the weekend ravers who wondered why the world they lived and worked in was so diametrically opposed to the one they played in
- the online junkies, the proprietors of fansites and moderators of lively newsgroups, the ones who got into the whole internet thing not because they saw dollar signs, not because they could program in C++, but because it was *new*, it was uncharted, it could be *theirs*
- the music fans who listened to *OK Computer,* listened to it over and over, listened and wondered, as did the aliens hovering high up above in "Subterranean Homesick Alien," wondered what was up with "all these weird creatures who lock up their spirits, drill holes in themselves, and live for their secrets"
- thousands of people who, yes, maybe didn't know what exactly the WTO did, couldn't give a coherent lecture on monetary policy, couldn't explain precisely what "most favoured nation" trading status would do to the price of tea in China or the quality of life in Tibet
- everyone who even now looks at it all and says: OK, yes, maybe Afghanistan needed to be bombed, maybe some of those people needed to be detained, surely the airports needed better security, but still it's not about *shopping;* I love my country and I value freedom and justice and innocent people should never have to die in an office building turned into an incinerator, but that *does not* mean everything was just fine before and that the only thing left to do is get back to normal, because normal was pointless and suffocating, and I want to know what it is we believe in, why, with all this money and power and knowledge, why have we not built the great society we claim to be defending?

Yes, what happened was that all of these people saw a crack, and they jumped at it. The point is not that the protestors in Seattle were right but that they were *engaged.* You can agree or disagree with them, with me, but *be engaged.* Care. Give a shit. If you're not engaged, then the worst people, the ones full of passionate intensity, will be the ones who map out our future.

I HAVE ASSEMBLED THESE DETAILS like threads, patches of fabric, formless pixels. I've put them together into a whole. It is wish fulfillment. I hope something will come together. *Is* coming together. I hope we've learned that prosperity is not the same thing as happiness, that novelty is not the same thing as progress, that connectivity is not the same thing as community. I hope I can use *The Simpsons* as my blueprint, hold it up against the changes that come in the next decade, the one after that, and see that each hypocrisy the show mocks has been addressed. I don't know what exactly needs to be done, and lord knows I'm not in charge. But I do know that I see debates on MetaFilter about the problems that are far more reasoned and compassionate than the ones on CNN. I know that I trust Jon Stewart infinitely more than Dan Rather. I know that Olivia Chow, by airing a single episode of *The Simpsons,* added about a half hour more insight into Canada's political discourse than the previous ten years of Question Periods combined. I could go on.

The thing is, I don't want to be able to assume beforehand that all politicians lie to sustain themselves, that all business leaders cheat to make more money, that all the news will be stupid and bad. Not anymore. I'm tired of it. Aren't you? I hope you are.

I could have sworn I saw a light coming on.

3

ANTHROPOCENE LANDSCAPES

From practically my first assignment meeting at *Shift*, I'd been agitating to write environmental stories, though I never did manage to sell the magazine on a single one other than a tangential rant about the digital revolution's climate change blind spot. It was only after writing a successful book about *The Simpsons* that I found editors willing to take gambles on my less digitally connected pitches, most of them stemming from research I was conducting in pursuit of two books about climate change and sustainability.

The time I'd spent tracking dotcom lunacy through the late 1990s served me well. By the time I hit the sustainability beat—which came with its own heavy doses of outsized ambition and breathless cleantech hype—I'd developed some skill at separating durable stories from passing fancies. I also knew how to navigate uncharted terrain and find unstated connections between disparate technological and cultural phenomena. I started in 2005 with a simple mission: to avoid fearmongering and apocalypse porn and instead document solutions, things that work, a future worth fighting for. This journey through a new kind of industrial age turned from a single book project into a career arc with no end in sight.

It seems too tidy somehow—too *scripted*—that the first decade of a new millennium would be caught between two discrete operating systems, but such is the fate of the world in the early years of the twenty-first century. There is the fossil-fuelled industrial age, and there is whatever we can invent to replace it before it drives us off a cliff. There's little in the way of stability or solid ground anywhere. From flailing post-communist dictatorships to the pristine coast of British Columbia to the seemingly timeless majesty of the Great Barrier Reef, all is now in flux as never before. Anthropocene means *wrought by human hands*, and like many handcrafted things this new epoch is beautiful, fragile, flawed and dangerously transient. Radical change is both an unfolding catastrophe and an emergent opportunity. It remains to be seen which side of the scale is the heavier one.

THE AGE OF BREATHING UNDERWATER

The Walrus, October 2009

Sometimes, as a writer of narrative non-fiction who is not exactly a journalist, you need to manufacture a reason to leap blindly toward a story.

I was in Australia for a series of lectures, as noted in the following story. I knew I had to go to the Great Barrier Reef, to see the place up close at least once in its pre-Anthropocene glory. I had no idea what I would make of it. My wife and daughter were with me, so I sold the Globe and Mail *on a travel piece about visiting the reef with kids. This gave me an excuse to beg for freebie hotel rooms in places I couldn't otherwise afford.*

The evening after my first dive on the reef, I sat at a faux-British pub in the tourist Potemkin Village that is Hamilton Island's harbourfront, and I started to scrawl random reflections and wonder about scuba as a metaphor. Somehow I convinced John Macfarlane at The Walrus *to commission 8,000 words with which to expand on those half-formed ideas. I turned in more than 10,000, and I nearly fell over when he asked for more. A generous and trusting editor is a gift.*

But here, in poise and in hard thought, I look down to find myself happy.
                              —William Matthews, "Skin Diving"

HARDY REEF, QUEENSLAND / 19°44' S, 149°12' E /
DEPTH: 10–13 METRES / JUNE 2008

HERE ON HARDY REEF, in soft, refracted light ten metres below the
surface of the Coral Sea, the scene is a sort of mystical mix of frenetic
and tranquil, like something out of an anthropomorphic cartoon
rendering of rush hour with the sound turned down to a soothing,
ambient hum. Dense aquatic traffic moves everywhere, constantly, in
swarms and tidy lines and tangents, in ever-morphing smears of electric colour, all against a backdrop even more riotously hued and just
as busy. The reef wall is a vertical forest of coral branches and squiggles in a hundred shades of Day-Glo, fire-red fractal twists alongside
lime-green paisley curls alongside delicate golden folds.

I slow myself, flapping my polyurethane fins just enough to stay
horizontal and giving a half-second push on the thumb-activated,
rubber-capped trigger at the end of an insulated tube protruding
from my buoyancy control vest. The vest inflates slightly with a gentle, echoing scrick. My breath leaves me in a burbling stream of rising, distended bubbles, a quiet bass drum rumble augmented by the
brushes-scraping-across-snare click of my demand regulator's valve.
Neutral buoyancy achieved, breathing free and easy beneath four

storeys of salt water, I hover an arm's length from the reef wall in an alien repose that feels much closer to floating magically above a forest canopy than to swimming.

"It is excusable," Charles Darwin wrote from the deck of the *Beagle* in 1836, "to grow enthusiastic over the infinite numbers of organic beings with which the sea of the tropics, so prodigal of life, teems." Word to that, Charlie—and you didn't even have the scuba gear to bring you down here to front-row seating so close to the prodigiousness you feel almost as if you're *of* it.

Hold still and fix your gaze anywhere, really, and the universe is born anew in otherworldly hues and you-gotta-be-kidding-me shapes. Tiny fish the colour of a roadworker's safety vest or the hat on a backwoods hunter's head—don't-hit-me yellow, don't-shoot-me orange—dart in and out of the coral fronds. The waving tentacles of an anemone have a multihued, translucent aspect, like something you'd find protruding from the forehead of a comic book alien. A school of trevally cruises by in an orderly procession so broad and dense you think for a second the sea is actually made of fish. Look down, and there's the mouth of a giant clam, the orifice mostly sealed by a membrane of electric purple and emerald green in intricate screensaver squiggles. Sergeant major fish will swim right up to the tempered glass of your mask and wait there expectantly like autograph seekers, giving you ample time to wonder whether you'd best describe the yellow splotches on the upper part of their bodies as *ballpark mustard* or *canary*, and to notice the way their distinctive black vertical stripes smear into a singular shade of pale neon blue at the dorsal fin and tail.

And then there's the extraordinary symbiotic web the reef's myriad denizens have woven, enabling this aquatic Babel to thrive more or less self-sufficiently for millennia. Hermaphrodites and sex changers abound. A great many of the reef's coral polyps mate once a year, simultaneously, in a great cloud of eggs and sperm whose release is precisely timed with the lunar cycle. Certain species of parrotfish, their scales resembling the most ambitious palette Matisse ever dared to play with, start out female and then switch gender in four stages, each more vividly coloured than the last, eventually reaching a dazzling final act that's been called "super-male." Some of the zaniest colouring elsewhere, meanwhile, is "aposematic," which is to say it's a warning flag to would-be predators of something that tastes terrible or hurts to bite.

Damselfish tend permanent gardens of algae and chase off interlopers like rabid guard dogs; blue tang have taught themselves to successfully invade these gardens in swarms. Different species of nocturnal squirrel fish have specialized within their genus on particular kinds of food—some eating only this type of crab, others only this size of prawn. Speaking of which, there are several species of prawn and certain kinds of small fish as well, collectively referred to as "cleaners," who live a significant chunk of their lives inside the mouths of other fish.

It's like contemplating the cosmos under a starry sky, only the planets and constellations are close enough to touch, and they swirl past you as though you were a comet in their midst. I could float here in this reverie until my oxygen ran out—divers have—but I'm eased out of my trance by the dive leader. She's a credentialed marine biologist, and she's just risen from a sea-floor bed of coral below us with a rare treat in her hand. She waves her hand up and away like she's releasing a bird, and I watch a flatworm the size and shape of a piece of Scotch tape drift slowly downward between us. The body undulates like a flag in a breeze as it descends, revealing a black sheen on one side, an explosion of deep red and purple and yellow on the other. The reef's endless variety in microcosm, twisting and dancing between us on the current. I read somewhere that they once analyzed a single three-kilogram chunk of dead Great Barrier Reef coral and found 1,300 worms from about 100 species living in its labyrinth of orifices and folds.

All due respect, Charlie, but it's not just that the reef teems with life; it's that it seems like the reason we coined the term. To *teem* is to exhibit the properties of a coral reef. I watch the flatworm descend against a writhing backdrop of coral fronds and bustling shoals of fish, and I try to imagine what it means that this little kidney-bean reef is one in a chain of 3,650—a profusion of life stretching from a few hundred kilometres south of here all the way up the remaining length of the northeastern Australian coast and beyond, 1,800 kilometres in total. The Great Barrier Reef: the largest living system on earth, a deeply interconnected macro-organism comprising nearly a tenth of all the coral reef there is.

Coral reefs worldwide occupy only about 0.17 percent of the ocean's surface area, but they provide habitat for nearly one-quarter of its marine life, all of it derived from a flawless, fragile symbiosis between coral polyps—the animals that form all those elaborate, plantlike structures—and a particular strain of algae called zooxanthellae. The polyps, fed by the photosynthesizing zooxanthellae that live on their bodies, grow up to secrete their own skeletons, bundles of calcium carbonate the zooxanthellae combine with ionized carbonates dissolved in the ocean's water to cement into place as limestone, thus providing a hospitable habitat for yet more corals.

This is a very delicate balance, and every dreamcoated fish and giddily tinted worm depends on it. Five times in the geological record, reefs have mostly or entirely vanished from the face of the earth, leaving corals to drift on the ocean currents for millions of years at a time, largely dying off until their preferred climatic conditions returned. Still, I find it impossible from my subaquatic vantage point to imagine that Hardy Reef could die, that virtually all life could one day cease to exist on the Great Barrier Reef as a whole. It's a fantastical notion, a witch's curse in a Teutonic fable. Maybe the sun won't rise tomorrow, and maybe the coral reef will no longer teem with fish. Surely such a dying would come here only at the very last, at the black end of some much wider catastrophe.

Well, sorry, Charlie, but here's a twist you couldn't possibly have seen coming from the crow's nest of the *Beagle*: It's the water. It's grown too warm, and absorbed way too much carbon dioxide. The pH is off, down from 8.2 to maybe 8.1. Not much, but it could mean everything. The beginning of some kind of unthinkable end.

I've only just heard. That's why I came.

MELBOURNE, VICTORIA / 37°49′ S, 144°57′ E / CAFÉ PATIO, NEWQUAY PROMENADE / THREE WEEKS EARLIER

THE AGE, JUNE 7, 2008:

Australia's Great Barrier Reef, more than 25 million years in the making, is "an icon primordial wilderness," says Dr. Veron—it is the greatest structure created by life on earth. The idea that it might be mortally threatened within the span of a generation or two he would once have considered preposterous.

"I was wrong," he says.

Twin assailants, both creatures of climate change, threaten the reef and oceans more generally. The lesser of these is the warming of the water, which turns the single-celled algae on which corals rely for their sustenance toxic, compelling the coral to expel them and probably die—the event known as coral bleaching—or to keep them and certainly die.

The worst bleaching events of history will become commonplace by 2030, says Dr. Veron, and by 2050, "the only corals left alive will be those in refuges on deep outer slopes of reefs. The rest will be unrecognisable—a bacterial slime, devoid of life."

The even greater threat is ocean acidification—the dissolving of carbon dioxide into the sea, forming weak carbonic acid. This is the climate change frontier to which science is swinging [with] increasing focus, as alarm grows at the threat it poses to marine ecosystems and to human food supplies and economies.

This Dr. Veron's byline in *A Reef in Time*, the chronicle of the life and impending death of the Great Barrier Reef whose release prompted this stunning report, reads "J. E. N. Veron," but the author actually goes by Charlie. That's Charlie as in Darwin—"Little Mr. Darwin," one of his schoolteachers once called him, because the prepubescent J. E. N. Veron was so obsessed with bugs and such. Somehow the nickname mutated into Charlie, and it stuck, even though Veron himself didn't read *On the Origin of Species* until years later. He studied dragonflies to earn the "Dr.," and for reasons I'll come back to he abandoned a promising career in entomology to become the world's most prolific coral taxonomist (having personally identified and named 23 percent of the planet's coral species), former chief scientist at the Australian Institute of Marine Science, and nowadays a *grise* whose *éminence* in his field is so uncontested that he is identified, in the press releases accompanying scientific declarations with dozens of Ph.D.-wielding signatories, as "the world's foremost expert on coral reefs."

So there I was, leafing idly through the pages of Melbourne's prestige daily, a cool winter breeze coming in off the harbour and setting its corners flapping, my mind still fuzzy with jet lag of International Date Line grade. A sunny morning on the pretty new boardwalk in the city's redeveloped Docklands, a cup of bold Melburnian coffee, one of those inside-page special interest stories to pass the time. A pair of full-colour before-and-after pictures of coral bleaching. And the world's foremost expert on coral reefs saying, without much in the way of qualification, that his learned opinion is that the Great Barrier Reef will be largely bereft of life by 2050, after which time whatever remains will be finished off by the over-acidified Coral Sea. That whatever world we might, if we're quite lucky and extremely bold, wrest from the jaws of this mutated climate, whatever equilibrium we might hope to reach, it is already not the one we knew. Not the one that provided the stable foundations for 12,000 years of this thing we call civilization. Not the Holocene epoch at all, perhaps, but something else, something new.

The term *Anthropocene* came quickly to mind. It translates loosely to "wrought by human hands," and I've used it with some frequency. I'd even stood before an audience at Melbourne's stylish new BMW Edge amphitheatre right there in the city's heart the night before and explained what I believed it meant. But reading *The Age* that morning—that was the first time I'd really *felt* it.

In time, it sent me searching for two other pieces of esoteric but existentially critical science news that emerged in 2008 and quickly vanished in the churning media sea of burst housing bubbles and flailing banks. The first was a concise policy paper released in August and signed by fourteen of the world's top ocean researchers—Charlie Veron among them—under the title *The Honolulu Declaration on Ocean Acidification and Reef Management*. Their statement noted the imminent onset of levels of ocean acidity not seen in "tens of millions of years," which would "compromise the long-term viability of coral reef ecosystems." The crisis was phrased more bluntly in the accompanying press release: "In July 2008, scientists at the International Coral Reef Symposium in Florida declared acidification as the largest and most significant threat that oceans face today and conveyed that coral reefs will be unable to survive the projected increases in ocean acidification."

The other crucial science story was a paper published in February 2008 in GSA *Today*, the Geological Society of America's house journal.

It was entitled "Are We Now Living in the Anthropocene?" and co-written by twenty-one members of the Stratigraphy Commission of the Geological Society of London, England, the body responsible for naming and dating geological time. "Sufficient evidence has emerged," went the closing argument, "of stratigraphically significant change (both elapsed and imminent) for recognition of the Anthropocene—currently a vivid yet informal metaphor of global environmental change—as a new geological epoch to be considered for formalization by international discussion." Which is to say the notion that humanity had permanently and fundamentally altered the planet was no longer a rhetorical flourish (as it had been when the chemist Paul Crutzen proposed the term "Anthropocene" a decade earlier) or a bit of activist hyperbole (as it had sometimes seemed coming from environmentalists a decade before that) but an emerging scientific reality.

I'm trying not to hyperbolize this. I don't want to traffic in visions of apocalypse. On the other hand, how can the probable demise of the ocean's most fecund ecosystem and the possible dawn of a new geological epoch be overstated?

Were this a traditional environmental narrative in the Thoreauvian vein, now would be the part of the story where I would shift summarily to lamentation. The tragedy is obvious, its scope impossibly huge, the loss beyond measure. But we have enough laments. More than our share. I wonder, actually, if it isn't the lamenting itself that led us here, at least in part.

Paul Crutzen traces the birth of the Anthropocene to the invention of the steam engine in the late eighteenth century, and modern environmentalism was born in that kettle as well. Thoreau's legendary trip to the wilderness of Walden was, after all, a reaction against all that the steam engine had wrought. For 150 years thereafter, an elaborate and often achingly lovely philosophy of the purpose of the human experiment held sway in green-minded circles, predicated on Thoreau's recommendation "to front only the essential facts of life," to go to the woods and "live deliberately." To become in some sense *part of the woods*, indistinguishable in action and impact from a fish or a frog or a fly. Finding a place of eternal harmony with Nature— ideally one sufficiently pristine to be worthy of Romanticism's awed capitalization of the word—became the goal of the environmental movement. The earth thrived in an exquisite equilibrium, and the enlightened seeker sought to find a human place within it.

If, however, that equilibrium is permanently altered and transmuting as never before, what do we even mean by "harmony"? What does it mean, moreover, when we know that this new order—whether we choose to call it Anthropocene or simply acknowledge its complete lack of historical precedent—was wrought in large measure by human hands? What is the proper goal of an Anthropocene conservation effort? How do we go about being sensitive to an Anthropocene ecology? What, in short, does it mean to be human in an ecological order of our own design?

I can't answer any of these questions definitively. These are early days, in turbulent weather. All new regimes are for now provisional by necessity. But I think I know where we should start. We need a new kind of story, a new template for our ecological philosophy—one that acknowledges what we have lost and the emerging limits of what can be saved, *but does not lament*. To borrow the terminology of the linguist George Lakoff, we must first change the frame.

The weight of a story is not just the sum of its details but also—maybe even primarily—a function of its structure, the way its plot points and archetypes map onto the ur-narratives (classical, Biblical, market triumphalist, what have you) that are deeply etched into our collective consciousness. A lament is by its nature nostalgic, downbeat, defeatist. It is predicated on a loss presumed to be absolute. Adventure stories, on the other hand—heroic narratives of victory against impossible odds in heretofore uncharted realms—these are the tools of transformative myth. This is what we need: a new myth of the frontier.

We are headed somewhere unknown, somewhere surely dangerous but also perhaps blessed with unexpected beauty. The terrain will be at least partially alien, the logic and rules of the place governed by inversions and seeming perversions of the natural order we've always known. Some of the tools we'll need to traverse this new landscape safely may at first appear unfamiliar, unwieldy, inconvenient. We may only comprehend their vital necessity once we've taken the plunge into this tumultuous sea. But we will learn to thrive. Feel exhilaration in the place of anxiety and lament. We will all learn to breathe underwater.

GLENOGLE, BRITISH COLUMBIA / 51°18' N, 116°49' W /
TRANS-CANADA HIGHWAY, EASTBOUND / AUGUST 2008

IN TERMS OF MYTHIC IMPORT and sheer physical presence, the Rocky Mountains are as close a Canadian analogue as you'll find to Australia's Great Barrier Reef, and they cast my little Honda in deep shadow as I drive east out of Golden toward the Alberta border. There's a new four-lane bridge over the Kicking Horse River a little way east of town, a staggering feat of modern engineering that has to be the tallest structure in the Canadian Rockies. Just on the far side of it, I find myself staring straight at a towering wall of rock covered in the rust-coloured skeletons of pine trees. The devastation on this slope appears total; the entire mountain face is a carpet of eerie orange-red. Around the next bend, the vast green blanket of forest is marred by blotches of the same sickly rust, as if stricken by some exotic rash or pox.

I can remember driving this stretch of highway only five years ago, and the green of the pines was everywhere and eternal, muted only by the winter snow. The mountain pine beetle, its population no longer sufficiently culled by winter cold to protect these forests from its wrath, has taken an Anthropocene toll so ruinous and rapidly exacted it's almost monumental. It could paralyze you to stare at it too long. You might decide there's not much more to be done, really, besides compose another lament.

Better not to dwell on it. But look just long enough, if you're still unconvinced, to recognize that this is not a regional conservation issue, not a problem limited to places of abundant coral life. Look just long enough—it shouldn't take more than a moment—to know in your bones that this is about much more than the quality of the scenery on your next vacation. That this, too, is the exotic undersea world where we must make our home. Look just that long. And then check the gauge on your tank, bite down hard on the mouthpiece, breathe slowly and steadily through your regulator, and carry on. If you linger too long, you could drown.

This is the scarred Canadian face of the problem, and it's a particularly tidy irony that some of the planet's most visible evidence of its proximate cause is embedded in the mountain rock just a little way farther east down the Trans-Canada. I crane my neck a bit to watch for mottled striations midway up the bare slopes of exposed rock—telltale evidence of the verdant tropical reefs that thrived here hundreds of millions of years ago.

Just before the border, I spy one particularly notable example, a rock formation known as the Burgess Shale. Discovered in 1909 near the base of Mount Burgess and excavated by a Smithsonian scientist named Walcott, the shale outcropping contained the remains of 65,000 different plants and animals—the best fossil snapshot ever discovered of the teeming life that thrived on earth between 530 and 542 million years ago, during an age so verdant it was dubbed the Cambrian Explosion.

The ancient tropical sea that is now Alberta was for many millions of years the site of a great many other reefs, built in those times by algae instead of coral polyps. And there was one in particular, beneath the soil of the town of Leduc down on the prairie, where the porous reef rock of the Late Devonian period trapped the remains of a staggering abundance of carbon-based life about 400 million years ago. Across the intervening millennia, those remains were slowly crushed into pools of thick black ooze buried a mile underground, and in February 1947 a drilling rig struck the largest pool anyone had ever tapped in Canada. Within five years, 137,000 barrels of crude oil were being extracted from the liquefied boneyards of the ancient reef every day, and the economy of Alberta has been driven by the search for energy-dense Devonian fossils ever since.

As well, it has become increasingly fashionable in this country to think of this as merely a regional issue, some kind of late-blooming prairie madness. But of course the produce of Alberta's Devonian bounty is a substance virtually all Canadians use in one form or another pretty much every day of their lives. If it is not in the tank of your car, then it is heating your house, or protecting your daily bread from mould, or keeping your body erect in place of the hip joint you were born with. And Alberta is just one relatively small outpost in an inexhaustible global campaign of seemingly limitless scale to extract ever more of the stuff, to burn it as quickly as possible. And thereby to release that ancient carbon in the form of carbon dioxide, in clouds of such prodigious volume, we've recently discovered, that they are well on their way toward extinguishing the abundant life from the reefs of today. Creating ready-made oil fields for whatever civilization comes to thrive 400 million years hence.

Understand: the heroic narratives of the Anthropocene are nothing if not ironic.

AIRLIE BEACH, QUEENSLAND / 20°16' S, 148°43' E / DEPTH: APPROX. 4 METRES / JULY 2008

THE OBVIOUS RESPONSE to the bonfire of the fossils that afflicts Alberta and everywhere else would be to impel those who are stoking it to cease and desist. Like, *now*. Stand before the majestic and plentiful landscape they are condemning to death and block the march of industry bodily. And so here, then, is what would appear to be an Anthropocene hero's pose: two divers in black neoprene floating defiantly astride a particularly photogenic patch of Great Barrier Reef coral, little vertical clouds of exhaled air extending from the tops of their heads like cartoon exclamation points, a neon yellow banner unfurled between them that reads KEEP THE REEF GREAT. And below that, in an instantly recognizable faux-scrawled font that has for a generation now denoted selfless action on behalf of Mother Earth: GREENPEACE.

The photo of this scene is dated July 22, 2008, and it is in many ways a textbook Holocene protest. And that's the problem. It's a symbolic act, staged with no intent other than to be photographed, disseminated, blipped to newspapers and websites around the world. Which publications may or may not publish it; which have over the years published a sufficient number of similar scenes of unfurled banners bearing strident messages composed in the imperative mood to render this one commonplace; and which in any case won't likely provide much in the way of context other than a blurb a few lines long about the proposal to drill into the shale seabed near this spot to extract oil, and the long-standing, vehement, and virtually unanimous disagreement of environmentalists with that kind of proposal.

They likely won't report any of the most salient details. That there is, for example, a flotilla of more than ninety vessels just four metres above the heads of these two beseeching divers, bobbing in solidarity in the Coral Sea's current. Nor that much of this solidarity comes from the nearby resort community of Airlie Beach, whose economic health depends—as does an estimated $5.8-billion share of the Australian economy—on the health of the Great Barrier Reef. And nowhere will it be reported that the reef is, in the opinion of the diver on the right in the photo, almost certainly not going to be kept great, regardless of whether oil drilling is permitted nearby. And most of all that the diver on the right is, yes, Charlie Veron, the world's foremost expert on coral reefs, and that until the moment he clutched the left side of that banner in his hands he'd spent a career of thirty-plus years as a marine scientist never once having participated in an environmental protest.

For Veron, the path to Greenpeace-branded activism began shortly after the publication in 2000 of his life's work, *Corals of the World*. Probably the most thorough taxonomy of the planet's corals ever assembled, it is a lavishly illustrated three-volume reference work cataloguing 800-plus species, almost 200 of them photographed by Veron himself. For his next project, he gave himself five years to research the big-picture story of the Great Barrier Reef—to move beyond coral taxonomy into the geological past and troubled climatic future of the natural wonder he'd spent thousands of hours studying at close range through the tempered glass of a dive mask.

"When I started writing my book," he later said on Australia's ABC Radio National, "I knew that climate change was likely to have serious consequences for coral reefs. But the big picture that emerged, quite frankly, left me shocked to the core. This really led to a period of personal anguish. I turned to specialists in many different fields of science to find anything that might suggest a fault in that big picture. I was depressingly unsuccessful."

Instead, Veron wrote the only book he felt he could honestly write. *A Reef in Time: The Great Barrier Reef from Beginning to End* is a densely detailed and often highly technical natural history of the reef ecosystem from prehistory to near-future. Veron assembles a thorough analysis of the health of reefs globally, using the Great Barrier Reef as his case in point, and examines the five mass extinctions that have afflicted coral reefs to date. As scientists must, he entertains a wide array of causal theories for these extinctions before coming to the conclusion that the deciding factor each time was the reduced pH level of the oceans, caused by absorption of increased concentrations of carbon dioxide in the air above them.

Nowadays, about a quarter of the carbon dioxide humanity emits each year is absorbed into the seas, where it forms weak carbonic acid and nudges down the ocean's pH, reducing the quantity of ionized carbon available for algae to fix along with coral skeletons as permanent limestone reefs. Thus bringing us, Veron argues, to the very precipice of the sixth mass extinction. When he takes up this possibility, his writing is propelled by a frankness and passionate urgency seldom seen in a science text. "Our tampering with the Earth's climate," he writes, resembles "a game of Russian roulette, with every chamber of the cylinder loaded. We must take the bullets out—all of them, quickly and at any cost—for this particular gun has a hair trigger and devastating firepower."

Veron's gravest concern is a tipping point in the ocean's chemistry that he calls "commitment," which comes freighted with the darkest of Anthropocene ironies. "Commitment," he writes, "embodies the concept of unstoppable inevitability, according to which the nature and health of future environments will be determined not by our actions at some future date but by what is happening today." In the case of ocean acidification, "the lag time of the ocean will make acidification a *fait accompli* before it has barely started." By the time we know for sure we've reached the tipping point, in other words, it will be far too late to alter the process.

Ocean acidification is in its infancy as a field of scientific inquiry. The term was only coined in 2003, and the first major research mission dedicated to its study was launched off the coast of Florida in late 2008. We are only beginning, that is, to discover the gravity of the crisis. As recently as 2005, scientists believed this commitment point (under a business-as-usual emissions scenario) was as far off as 2050, but many now date it to about 2030. Veron, ever the iconoclast, believes our current emissions trajectory could reach commitment "within a decade."

I wish I could report a note of caution or qualification from my own conversations with him. A few days before Christmas last year, he sent me a grim sort of season's greeting by email: "I have just come back from a three-week trip to the far northern GBR—although it seems impossible I really think the whole place is going to die before I do. Rather takes the edge off diving."

I reached him by phone a while after that. "Well, I'm sixty-four years old," he explained, "and the probability is that if I remain healthy and active I will see this process in my lifetime. I've got two young children, and they'll certainly see the end of the Great Barrier Reef if they live a normal life. That's how immediate this is. There is no doubt about it. If there was a doubt, boy, I would jump on it, and so would a lot of other scientists. The thing is, though, that there isn't any doubt."

I should stress that this is the conviction of a man who has probably spent more time in direct communion with coral reefs than any other human being who has ever lived. Something like 7,000 hours, by his own estimation—the equivalent, in terms of more common careers, of more than three full years of forty-hour work-weeks. Breathing underwater, immersed in the teeming life of the reef. I think I can safely state that there is no one on earth more fully invested in its survival.

This is why there's a certain kind of uncommon heroism in the simple fact that Veron's response to this existential threat to his spiritual home has been to commit himself wholeheartedly to the act of trying to save it. It's a quixotic if not wholly futile mission, according to his own well-founded scientific convictions. To know better than anyone the vanishing slimness of your chances and to take action anyway—Sisyphus would recognize that choice. It is an act of faith in the human spirit, its ability to adapt and renew, one as wholehearted and profound as any I've known.

It could be the cornerstone, all by itself, of an enduring Anthropocene ecological consciousness—to act knowing much of the loss has already been tabulated, because not to act would be to deny the basic right of humanity to its own survival, to deny our children the right to their dreams. "I'm not trying now to talk up saving the Great Barrier Reef, because I think it's a lost cause"—this is how Veron puts it. "I am talking up: Look at the Great Barrier Reef and think about it. Do we want the rest of this planet to follow suit?" Still, lost cause or not, the Great Barrier Reef remains dear enough to him that he couldn't stand idly by and let it be defaced for no good reason. And so it's understandable that he would react with utter contempt upon learning of the plan to bring industrial oil drilling equipment to the Great Barrier Reef—"an exceedingly dumb thing," in his estimation—and that'd he'd leap at the chance to unfurl a banner in its defence, as conventional an act as that is.

The most salient thing about his protest for me, though—the detail that points the way toward a sturdier new frame for Anthropocene ecological action—is not the banner or the reef, but the apparatus that permitted Veron to breathe freely as he posed there. The scuba gear, I mean.

This is indeed one of the most curious and surprising things about Veron's career: after he finished his Ph.D. in the neurophysiology of dragonflies back in the early 1970s, his sole qualification when he applied for a vacant post as a coral researcher at James Cook University on the Queensland coast was that he was a certified scuba diver. Had he not become a recreational diver while he was a student down south near Sydney, he'd never have found his way to the Great Barrier Reef.

What's more, he became pretty much the first coral taxonomist to conduct extensive underwater field research. The whole body of knowledge on the nature of corals and the reefs they build was to that point based on the lab-based study of specimens. Veron's scuba-aided work quickly dispelled a whole range of faulty theories. We simply wouldn't know how the reef works as a living system, nor the true nature of the trouble it's in, if Charlie Veron hadn't learned to breathe underwater.

Another Anthropocene irony: neither Veron nor anyone else would have been able to breathe underwater in the first place were it not for the whole smog-belching mess of modern industry. There are indeed few things a human being can do that are as wholly industrial—as totally dependent on the development of an industrial society so complex and muscular it has changed the very chemistry of the seas—as scuba diving.

This is why, to my mind, Veron's underwater protest got the framing all wrong. We're not trying to keep the reef great, and his own writing attests to the fact that we're past saving the reef in anything like its present condition. We're trying, actually, to save scuba. To conserve a place on this planet for nearly seven billion people and counting, and for the whole baroque industrial society that permitted us to swell to such ranks, and to produce industrial quantities of the rubber and tempered glass and polyurethane and neoprene that have enabled us to come to know the reef firsthand.

This is no semantic exercise. This is a fundamental redress of environmentalism's assumptions and priorities—of its basic frame. A focal shift from the pristine and natural to the unnatural and artificial, from salvation and conservation to self-preservation, from primordial ecosystems to modern industry. And so it might as well begin with the profoundly unnatural, self-preserving, industrial-grade act of scuba diving itself.

SANARY-SUR-MER, FRANCE / 43°7' N, 5°48' E / DEPTH: VARIABLE / JUNE 1943

IN THE WINTER OF 1942, a French engineer named Emile Gagnan developed a special demand valve that would allow motor vehicles to run on cooking gas. Conventional gasoline had become nearly impossible to obtain in Nazi-occupied France, and Gagnan's valve was designed to automatically feed cooking gas into a car's motor at the steady rate it required to run properly. When his friend Jacques Cousteau came to him around the same time, coincidentally looking for assistance with a similar problem regarding airflow for an underwater diving system, he adapted his cooking-gas valve to solve it. Gagnan's demand regulator—a valve held in place on the diver's face by a mouthpiece—was the final, essential piece of equipment needed to permit safe, effortless breathing underwater. Gagnan shipped a complete "aqualung" system to Cousteau's home on the French Riviera the following June, and Cousteau and his small cohort of diving pioneers logged more than 500 dives in the picturesque coves surrounding Sanary-sur-Mer that summer alone. In the summer of 1943, in the shadow of catastrophe, they invented scuba diving.

The history of diving is a kind of shadow history of the industrial age, mostly incidental to the primary arc of the story but wholly dependent on it, culminating in the serendipitous development of the demand regulator out of wartime motoring necessity. The first diving suits were produced in the 1850s from canvas rendered waterproof by a layer of mixed rubber and naphtha (a by-product of oil refining). The breakthrough in medical science that would eventually make diving safe—the discovery of the cause of the bends and how to prevent them—came a quarter century later, during the construction of the Brooklyn Bridge. The bridge's workers toiled on its foundations in "caissons"—giant, bottomless wooden boxes sunk into the bed of the East River and pumped full of compressed air to hold back the water. Several workers died and dozens more were gravely injured before doctors figured out that their patients needed to readjust more slowly to the lower air pressure at the surface to avoid decompression sickness. Another half-century later, in 1930, neoprene, the preferred material for modern wetsuits, became the first synthetic rubber ever invented when DuPont developed it for use in automobile gaskets and hoses. And so on, from the tempered glass of the dive mask to the polyurethane in the stiff, distended toe of the standard diving fin. Scuba's raw materials were the by-products and afterthoughts of the great industrial powers' quest for bigger and better bridges and cars and war machines.

In the towering shadows cast by this ever-expanding industrial order, another parallel history emerged. It began, more or less, with Henry David Thoreau's sojourn at Walden Pond—a deliberate, principled rejection of industrial society as a whole. Contemporary environmentalism traces a direct line of descent back through Greenpeace, Rachel Carson and John Muir to Thoreau's handmade cabin in the Massachusetts woodlot owned by fellow Transcendentalist Ralph Waldo Emerson, and among its strongest links to Walden has been an enduring anti-industrial bias. With good cause, of course. The culprit in almost every environmental crisis the movement has tackled—from the despoiling of animal habitat to Carson's DDT-poisoned birds to reef-killing greenhouse gas emissions—has been the resource-devouring, waste-belching march of modern industry. Thoreau's cause was righteous, his critique of industrial society trenchant, and it'd be no stretch to argue that the acidification of the world's oceans is a sort of ultimate proof of *Walden*'s profound, prophetic truths.

If, however, the goal of environmentalism in the Thoreauvian tradition was to halt the march of rapacious industry, that same pH imbalance might best be understood as the litmus test of the movement's failure. After more than a century of advocacy and action employing Thoreau's frame, the earth's natural wonders have never been closer to collapse. "Insanity is doing the same thing over and over again, but expecting different results"—so goes a widely cited aphorism attributed alternately to Ben Franklin, Albert Einstein, and Chinese proverb (it appears actually to originate with the American mystery novelist Rita Mae Brown). What if, instead, we followed the lead of another nineteenth-century prophet, a writer whose work predicted the advent of an eye-popping world of air conditioning, space travel, the helicopter, and—most famously—the untrammelled exploration of the subaquatic realm?

Jules Verne's *20,000 Leagues Under the Sea* was published in 1870, just sixteen years after Thoreau's Walden diary. In one particularly prescient passage, Verne's Captain Nemo describes to Professor Aronnax a new kind of "diving apparatus"—one freed of the rubber hose attaching it to a ship, instead employing an iron tank of compressed air worn like a backpack, feeding the diver through a mouthpiece fitted with a tongue-operated regulator. The diver would then use a sort of sodium-carbon lamp (analogous to sodium gas lamps not produced commercially until the 1930s) to light his way. "Thus provided," Nemo informs the enthralled Aronnax, "I can breathe and I can see."

In *20,000 Leagues* and other fantastical tales, Verne imagined a glittering world of wonders made possible by modern science and its industrial adjuncts. In so doing, he created the literary genre of science fiction, which, despite its penchant for dystopian scenarios, has proven, in the aggregate, to be at least as rich a repository of hope for an enlightened human future as Thoreauvian environmentalism has. And certainly the world most of us live in today looks much more like a Jules Verne novel than a primordial New England forest.

And so an Anthropocene ecological consciousness, emerging from the daily more manifest observation that the natural world has become as much a product of human invention as submarines and flying machines, might—it follows—be more in keeping with Verne than with Thoreau. It might, for starters, pick up a thread common to almost all science fiction: the idea that our future will by necessity be very, very different both from today and from some idealized pastoral past, likely much more artificial, and yet luminous in the constellations of possibility it offers.

There is indeed a whole emerging school of futurist thinking inspired by this aspect of sci-fi, one particularly prevalent in the pronouncements of digital-communications prophets. Sci-fi, so this line of reasoning posits, is the last real "literature of ideas," because it alone allows sufficient room for speculation on the limitless possibilities created by modern technology. Sci-fi-inspired futurism is not inherently materialistic: it is not about gadgets but rather about the staggeringly broad options for new ways of life being created in the digital age. In recent years, the Verne school has come to commingle with certain strains of post-Thoreau environmentalism—ones rooted mainly in urban spaces, interested as much in city planning and the energy economy as in whales and rainforests. And in that cross-pollination, we've begun to see the scaffolding of a new frame.

"Hairshirt-green is the simple-minded inverse of 20th-century consumerism," argues sci-fi author and futurist Bruce Sterling. "Like the New Age mystic echo of Judeo-Christianity, hairshirt-green simply changes the polarity of the dominant culture, without truly challenging it in any effective way. It doesn't do or say anything conceptually novel—nor is it practical, or a working path to a better life." In place of these Thoreauvian hairshirts, Sterling argues for a new frame constructed on the foundation of "sustainability," broadly defined as that which "navigate[s] successfully through time and space." Like certain of Jules Verne's amazing machines, perhaps. Or like any given diver on any old reef, breathing freely underwater.

Or like that exhilarated fatigue I've always felt at the end of a dive—a sort of quiet triumph, a sense of wonderment at having gone somewhere so far from the workaday norm and passed through without incident, emerging finally, reborn, to see the world with anointed eyes.

KOH PHI PHI, THAILAND / 7°46' N, 98°47' E / DEPTH: 16 METRES / JANUARY 2006

THE RECORD IN MY LOG of my fourth open-water dive reports my time at depth, breathing underwater, at just forty-four minutes. I find that hard to reconcile with the vividness and variety of the memories. We were diving a fringing reef not far from the cluttered resort island of Koh Phi Phi in southern Thailand. There were four of us on the dive—three students plus our instructor—and we'd intersected with another foursome from our boat down at the base of the reef wall to form a rapt, floating semicircle around a slumbering leopard shark for what felt like half an hour all by itself. We'd also taken advantage of the seeming weightlessness of neutral buoyancy to stage a mock punch-up in the style of *The Matrix* for the dive shop's digital underwater video camera. And then our instructor had taken us, finally, on the one part of the dive he'd told us ahead of time we shouldn't note in our logs. Because we weren't yet licensed open-water divers, we weren't technically allowed to navigate a coral cave.

He led us up the coral wall to a jagged oblong opening midway up the face. The cave was shadowed and dense with what seemed like an impenetrable mesh wall of inch-long silver fish. I remember only a twinge of anxiety in the wave of dawning exhilaration. The tiny fish, I'd later learn, are called silversides, and as they bobbed in the current their bodies would flip in undulating waves from shadowy blue-black to brilliant silver when they caught the mottled light from the surface. They formed a shimmering curtain at the cave's mouth, and they parted as I entered, as if responding to some automated infrared sensor, the vast shoal forming a perfect, swirling outline of my body to let me pass. They must have numbered in the thousands. For a long time, researchers believed that schools of fish had leaders, that each individual was responding to some chain of near-instantaneous cues, like a precision drill team. But more recently, they've come to understand that in anything but single digits the fish form a kind of unified collective, "more like a single organism than a collection of individuals." (The researcher Brian L. Partridge was specifically describing a school of silversides when he wrote that.)

For some indeterminate, fleeting moment that stretched out toward eternity, I was part of their unity. I was completely enveloped in them, a second skin of wriggling fish separated from my own by a few centimetres of sea water, nary a single one so much as brushing against my flapping fins. Wrapped in neoprene, freighted under a tank of some heavy steel alloy, ten metres below the nearest breath of fresh air—poised, that is, in a position beyond human experience for all but the past few decades of our existence—I felt a moment of communion with nature as close to total as I've ever had, and as transcendent as I could ever hope for.

I returned from my sacred rite with the silversides to the patio of a concrete tourist bungalow along the dense waterfront pathway of Koh Phi Phi, where I watched the sun set in a gentle orange sigh as the bars and clubs up and down the beach growled to life. Not even a tenth of the average night's tourist population could fit aboard all the island's dive boats, and no doubt only a tiny minority had bothered with the reef. One fundamental goal of the past 150 years of industrial age environmentalism has been to convey somehow a sense of wonder and respect for the natural world—the same soulful quietude Thoreau discovered after many months in the woods—to those who have lost or not yet found it. To raise the collective consciousness, thus to alter and finally invert forever our sense of humanity's place in the natural hierarchy—no longer seeking dominion and depletion, but instead pursuing rehabilitation, preservation, harmony.

Breathing underwater, in my experience, can provide a kind of hot-wired, instant-karmic version of this mystical oneness. But even this near at hand, it remained an afterthought. The reef's demise would mean seemingly nothing to the thrumming life of the island. There might be a way to move the bulk of humanity to act for the reef's salvation—and our own—but spiritual union with the reef, with nature itself, appeared to be too small and precise a tool.

HAMILTON ISLAND, QUEENSLAND / 20°21' S, 148°57' E / MARINA TAVERN / JUNE 2008

THE JUXTAPOSITION BETWEEN the reef's majesty and the workaday tourist buzz of my port of call was even more jarring on the Great Barrier Reef. I was staying on Hamilton Island, the most heavily developed resort destination in the Whitsunday chain, featuring the most easily navigated path to the reef. And now I sat on a pub's terrace watching another magnificent tropical sunset, this time as it descended over a marina crowded with yachts and pleasure cruisers.

At dusk, the small island adjacent to Hamilton was a purpling hump on the horizon, and I spied the silhouettes of construction cranes and bulldozers fixed along its spine like cubist remora. Until recently a fairly rustic little holiday spot, Hamilton Island was the site of a massive expansion beginning in the mid-1990s. Because the entirety of the Whitsunday island chain falls under the rubric of the Great Barrier Reef Marine Park Authority, however, Hamilton had reached its permitted development limit. The resort's new eighteen-hole golf course was thus being built on the little island across the channel from the marina.

For a moment, I was livid. I was barely an hour's journey from the planet's pinnacle natural achievement, its ancient wisdom and unsurpassed beauty nowhere more readily accessible than right here. What atrophied mind turned instead, in such a place, to the presumed necessity of a round of golf? The whole of the reef was in very real and proximate danger of vanishing forever. Surely before we came to such a precipice, there'd first be no more goddamn *golf.*

I was well into the next pint before my hubris caught up with me. Trace the outline of the extraordinarily complex apparatus that was required to bring me to my most recent holy communion with fan corals and neon-coloured fish, and try to imagine a thing more thoroughly dependent on the industrial order that was on the verge of sealing the reef's fate. The steel pontoon dock permanently moored to Hardy Reef, outfitted for scuba and snorkelling, hosting reef tours by semi-submersible and a buffet lunch for 200. The high-speed, wave-piercing catamaran tethered to the pontoon, nearly forty metres long and digitally stabilized for my comfort. The departure dock here on Hamilton Island—just one of a dozen in the tidy, modern marina, this little semicircle backed by high-end retail, purveyors of multinational cuisine, and the tavern serving up these cold draft beers imported from the mainland. The luxury condos and beachfront high-rise hotels scattered farther afield. The airstrip out on the peninsula in the distance, with daily non-stops to Brisbane and Sydney.

I felt as if I were balanced precariously atop this whole ungainly apparatus, and I recognized in that moment that it was the apparatus itself, not the beauty it allows us to touch, that we had to work to preserve. This whole affluent, hyper-advanced world order, which has given rise to a tourist infrastructure with gears so well greased it enabled me, basically on impulse, to arrange a day trip to the Great

Barrier Reef with little more effort than it takes to order a takeout pizza—or to make one from prefab groceries.

CRYSTAL WATERS, QUEENSLAND / 26°47' S, 152°43' E / AROUND BACK OF THE COMMUNITY KITCHEN / JUNE 2008

A FEW DAYS LATER, I went with my family to stay at Crystal Waters, a "permaculture village" on the banks of the Mary River near the Glass House Mountains of southern Queensland. Crystal Waters was one of the earliest experiments in sustainable living, founded in the late 1980s as a new kind of rural community based on the permaculture design concept developed by Australian ecologists David Holmgren and Bill Mollison. At its core, permaculture envisions human settlement as a kind of biomimicry—"the conscious design and maintenance of agriculturally productive ecosystems which have the diversity, stability, and resilience of natural ecosystems," as the Permaculture Research Institute of Australia puts it.

Crystal Waters is a lushly vegetated and loosely populated township of eightysome homesteads that spills across a series of steep hillsides and deep valleys. It turned out we'd arrived the day before its twentieth birthday party, and a crowd of current and former residents gathered the next morning on the village green for the celebration of what remains one of the most ambitious sustainability experiments in the world. We made our way down from the guest cabin to help out with the preparations. I joined a cluster of residents at the cob oven—a wood-burning stove handmade from straw and clay—to help out with the pizza-making operation. It had been my understanding that Crystal Waters was striving for self-sufficiency, so I was a little surprised to see a stack of prefab, plastic-wrapped pizza shells on a table next to the oven, along with a range of store-bought toppings. The nearest full-service grocery store was in a town almost thirty kilometres up the highway. Yet another Anthropocene irony: even the most ambitious sustainability pioneers are thoroughly entrenched in the order they're trying to supplant.

We fell into casual conversation in that easy way men do when there's a fire to be tended, and talk turned to my recent communion with the Great Barrier Reef. One of the locals, a maker of elegant slide didgeridoos of his own design, said something about how the future didn't look very bright for the reef.

"Not unless we really make some changes, no," I replied, shooting for a note of pragmatic optimism.

"Nah," chimed in a ponytailed dude named Angus who was assembling pizzas. "It's gone, mate. Might as well start getting used to the idea."

I didn't know what to say to that, so I busied myself with the cheese grater. His tone had sent me reeling, for reasons I couldn't place until much later. It wasn't grave or accusatory, not glib nor gleefully nihilistic. It was a win-some-lose-some tone, a shooting-the-breeze-around-the-bonfire tone. The tone of someone who'd already reached some sort of difficult reconciliation a good while back with the notion that there was nothing so sacred or durable that it exists beyond the reach of this tumult. It was the tone, I guess, of someone who'd dedicated his life to the step that came after the lament. I came in time to think of it as of a piece with future-tense voices like Bruce Sterling's—the emerging voice of Anthropocene reason.

There is a double edge, however, to this kind of realism. I've come to embrace it because it places the focus where it must stay—on the task at hand and the road ahead—but it also veers perilously close to a kind of bleak fatalism that invites self-serving rationalizations of inaction. It could even seem to encourage much darker survival strategies.

When I spoke, for example, to Rod Salm, the director of coral reef conservation at the Nature Conservancy of Hawaii, whose office convened the workshop that led to the Honolulu Declaration, he told me about an urgent call he'd recently received from someone on the board of trustees of a charitable foundation that provides significant funding for the conservancy. The board members had recently seen a study suggesting little hope remained for the world's reefs, and they wondered why, if the reefs were inevitably headed to hell in a handcart, they should invest more money. "We have to leave people with the encouragement that there are actions we can take, and if we do take actions, there is hope," Salm told me. "If we kick back and say there's nothing we can do about it, definitely we are going to lose things."

The Anthropocene epoch—the term itself is predicated on the idea of a permanent and categorical shift in the earth's equilibrium. Some of the most strident opposition to its use that I've heard, in fact, has come from the most passionate of green activists and sustainability advocates, who fear that it does nothing but yank the lid right off the most dangerous Pandora's box in the industrial closet—the one labelled "geoengineering."

Geoengineering—large-scale intentional tinkering with the planet's climate—is, to be sure, a uniquely hubristic school of applied doomsday science. And it is predicated on the twisted logic that a reasonable response to evidence that human industry has irrevocably altered the biosphere would be to undertake to alter it in much more intentional and grandiose ways. I'd argue, though, that while the fear of geoengineering is well grounded—the Strangeloves of the world are indeed lustily roused by the notion of irreversible climate change—the lid is wide open already.

Consider, for example, an article from the March/April 2009 issue of the journal *Foreign Affairs*. The title is "The Geoengineering Option: A Last Resort Against Global Warming?" A perfectly reasonable question mark, floating just above the names of five esteemed academics at A-list institutions (Stanford University's Law School, Carnegie Mellon's Department of Engineering and Public Policy, the Center for International and Security Studies at Maryland). Below that, a dispassionate analysis. All this in the same pages where the free world's Cold War containment strategy was first articulated.

"The time has come," the authors argue, to give serious attention to geoengineering—both because it might provide "a useful defense" against the worst shocks of a rapidly changing climate, and because some rogue state's "unilateral geoengineering project" could be launched before the repercussions are fully understood. There are any number of potential projects, they note, from seeding the lower atmosphere with sea water to create a reflective layer of dense cloud to launching clusters of giant reflective discs into orbit.

The most feasible and cost-effective geoengineering strategy, the authors explain, would be "launching reflective materials into the upper stratosphere," in conscious imitation of the 1991 eruption of Mount Pinatubo, which created a plume of sulfur dioxide particles large and thick enough to reduce the entire planet's mean temperature by half a degree Celsius for a time. Should humanity find itself in need of such a cloud to stave off global warming long enough to develop a more effective long-term solution, what would the costs and benefits be? As the authors note, it's really pretty straightforward, from a technical standpoint, to use "high-flying aircraft, naval guns, or giant balloons" to build a planetary-scale sunblock, whether made of sulfur dioxide or—let's get serious about this—"self-levitating and self-orienting designer particles engineered to migrate to the Polar Regions and remain in place for long periods." The real question is whether or not that would be a useful thing for us to do.

It warrants special mention that "The Geoengineering Option" is the first *Foreign Affairs* article ever to employ the term *ocean acidification*. And the authors duly note that no geoengineering scheme yet extant would do a thing to slow the acidification process, since blocking the sun does nothing to reduce the amount of carbon dioxide available in the atmosphere for absorption by the sea. "Fiddling with the climate to fix the climate strikes most people as a shockingly bad idea," they finally concede. But it's only prudent, they argue, to study geoengineering as "a true option of last resort."

Jules Verne would probably recognize this terrain, and his fellow sci-fi founding father H. G. Wells—who dreamed up Martian apocalypse and that mad scientist Moreau—surely would. Perhaps there are no broad new vistas of human ambition entirely free of dark clouds, and in any case geoengineering is one that will necessarily shade the blue-sky thinking of the Anthropocene. I'm convinced it's worth that risk, though. As the sustainability guru Paul Hawken recently told the graduating class of the University of Portland, "Civilization needs a new operating system, you are the programmers, and we need it within a few decades." Yet I can't see how we motivate ourselves toward an engineering project of that scale without stepping right to the precipice of total system failure and staring directly into that abyss—and then looking up at the clear sky above and understanding, in the same way we understand that it is a collection of gases that gives us life, that it is an Anthropocene sky that feeds us now.

"If you look at the science about what is happening on earth and aren't pessimistic"—Hawken again—"you don't understand the data. But if you meet the people who are working to restore this earth and the lives of the poor, and you aren't optimistic, you haven't got a pulse."

BALLARAT, VICTORIA / 37°37' S, 143°53' E / CARO CONVENTION CENTRE AUDITORIUM, UNIVERSITY OF BALLARAT / JUNE 2008

I FIRST HEARD TELL of "resilience"—not as a simple descriptive term but as the cornerstone of an entire ecological philosophy—just a couple of days before I met Charlie Veron on the pages of Melbourne's most respected newspaper. I was onstage for the opening session of the Alfred Deakin Innovation Lectures in an auditorium at the University of Ballarat at the time. The evening had begun with a literal lament—a grieving folk song performed by an aboriginal musician. I'd then presented a slide show of what I considered to be the rough contours of an Anthropocene map of hope, after which a gentleman I'd just met, a research fellow at Australia's prestigious Commonwealth Scientific and Industrial Research Organisation named Brian Walker, placed my work in the broader context of resilience theory.

I had to follow Veron all the way to the edge of the abyss his research had uncovered before I could come back around to resilience. The concept, it turns out, emerged from the research of a Canadian-born academic named Buzz Holling at the University of Florida, and has since been expanded by a global research network called the Resilience Alliance. "Ecosystem resilience"—this in the Resilience Alliance website's definition—"is the capacity of an ecosystem to tolerate disturbance without collapsing into a qualitatively different state that is controlled by a different set of processes. A resilient ecosystem can withstand shocks and rebuild itself when necessary." It's a concept I encountered repeatedly in my conversations with reef researchers.

Charlie Veron's *cri de coeur*—alongside several others—begat the Honolulu Declaration, which in turn provoked many reef scientists to radically redirect their work with resilience in mind. For example, they've started studying the coral polyps of the Red Sea, which thrive in water warm enough to kill most other reef-building corals, and have initiated investigations into the feasibility of a kinder, gentler geoengineering, in which more resilient corals might be seeded on ailing reefs.

When I spoke to University of British Columbia climate scientist Simon Donner, he was just back from the remote South Pacific island nation of Kiribati, whose fringing reefs have prospered despite taking the hardest hit of any in the world from the dramatic climate shifts of El Niño. "Let's target what we protect based on what has the best chance to survive—what might make it through some of the warmer temperatures," he explained. "Even if we freeze emissions today, the planet will continue to warm for a few decades. We have no choice but to do our best to help reefs adapt to at least some of our warming."

The Nature Conservancy's Rod Salm, meanwhile, was searching for ways to use old-school conservation techniques to encourage resilience by reducing the other stresses on the reefs. Significant damage from mass bleaching and acidification was now inevitable, but if the reefs afflicted by these catastrophes are in otherwise good health—not overfished, not poisoned by toxic effluent or overrun with the algae that prosper in the nitrogen-rich runoff from industrial farms—then they may survive long enough to benefit from the drastic reduction in greenhouse gas emissions essential to the whole planet's health. "Investing in stopping change is not a useful strategy," Salm told me.

This points to the broader implications of the resilience concept— the stuff Brian Walker likes to talk about. He and his colleagues in the Resilience Alliance often refer to their field of study as "social-ecological resilience," suggesting that people are as essential to the process as reefs or any other ecosystem, and that real resilience is created in the complex, unpredictable interplay between systems. "With resilience," Walker told me, "not only do we acknowledge uncertainty, but we kind of embrace uncertainty. And we try to say that the minute you get too certain, as if you know what the answer is, you're likely to come unstuck. You need slack in the system. You need to have the messiness that enables self-organization in the system in ways that are not predictable. The best goal is to try to build a general resilience. Things like having strong connectivity, but also some modularity in the system so it's not all highly connected everywhere. And lots of diversity."

Resilience, then, embraces change as the natural state of being on earth. It values adaptation over stasis, diffuse systems over centralized ones, loosely interconnected webs over strict hierarchies. If the Anthropocene is the ecological base condition of twenty-first-century life and sustainability is the goal, or bottom line, of a human society within that chaotic ecology, then resilience might be best understood as the operating system Paul Hawken was on about—one with an architecture that encourages sustainability in this rapidly changing epoch.

This new operating system will, by necessity, be comfortable with loss. There is, after all, much to be gained from epochal, transformative change. In the midst of chaos and devastation on the scale of a world war, for example, we might discover how to breathe underwater.

QUEANBEYAN, NEW SOUTH WALES / 35°20′ S, 149°15′ E / HEAD-QUARTERS, DYESOL / JUNE 2008

IN A SUBURB OF CANBERRA, Australia's rather sterile capital, in an industrial park populated mainly by nondescript aluminum sheds housing welding shops and auto brake installers—the polishers of the great industrial machine's many gears—I discovered what I've come to think of as a sort of harbinger of human-scale resilience. A particularly slick piece of software to run on that new operating system, if you will. The core of the Anthropocene crisis is energy, and what we do to the chemistry of the skies and the seas in order to produce it on an industrial scale, and in this desultory little industrial park I found an Anthropocene power plant.

In the gloom of a conference room at a company called Dyesol, with the blinds drawn and the lights off, I watched a new kind of solar panel send a bank of fans busily spinning, generating electricity from nothing but shadowed ambient light. The company calls its little power plant a Dye Solar Cell. It is a microscopically thin layer of dye made from a highly conductive metal called ruthenium, spread over the extraordinarily large surface area created by a base layer of titania paste.

The sample panels I was shown were smaller than the notebook page on which I sketched my description of them, and if all goes according to plan I will never see them mounted on great metal braces that protrude from a wide building's flat top or fill some sunny field end to end. Instead, they will essentially be printed onto sheets of steel as they roll along an assembly line at three to five metres per second. A company called Corus (formerly British Steel) in faraway Wales has already begun converting a production line used to manufacture its ColorCoat steel roofing. It produces 100 million square metres of the stuff each year—enough to re-roof every Walmart in North America—and as of 2012 it intends to start selling industrial quantities of its roofing with Dyesol's solar cells built right into it.

Australia's long-standing expertise in solar research began with tracking stations built for the American space program in the 1960s, some of them placed so deep in the outback it made technical and economic sense to go with costly, experimental solar power instead of running electrical lines from the nearest human settlement. A generation later—as a time-delayed by-product of maybe the most grandiose of industrial society's many ambitions—technology has emerged to crown the world's nondescript warehouses and hulking factories and sprawling big box stores with a thin skin of solar cells.

Look. Imagine one of those buildings, stage lit by the sun under a sky of clear blue, untethered at last from the smog-spewing hull of industrial society. This is what it will look like when a building learns to breathe underwater. This is the next frontier.

THE NEW GRAND TOUR

The Walrus, May 2010

During the research for my 2007 book The Geography of Hope, *I stopped off in Berlin as a way station between Copenhagen and the southern German city of Freiburg. I'd been looking at wind energy innovations to the north and was bound for solar industry breakthroughs farther south. By coincidence, an old Shift colleague, Ian Connacher, was shooting documentary footage for a few days in northern Germany, and he'd rented a flat in Berlin. I decided to stay longer than I'd planned—Berlin, after all, is always worth another day—and so I scheduled some meetings with federal government officials who might know a little bit about the German solar boom I'd come to believe was centred down in Freiburg.*

I didn't expect the canned presentations of communications people in Germany's international trade department to be scintillating; they were. I didn't expect to be as curious as I was about something one of them said toward the end. Something about the real hub of the solar boom in the old East Germany. Something about how I should take some more time here, check it out. It took three years, but when I finally made it back, it gave a theme to this piece and a thesis to my next book, The Leap. *Great stories are often happy accidents.*

Greetings from... Your Brightest Possible Future

It's a real pity, Gesine Bänfer wanted us to know, that we couldn't have visited her daughter's kindergarten at the end of last week. This scene was fine and all, she conceded: The building a low-slung, primary-coloured series of attached, self-contained classroom units, each with its own patio and garden, like a row of holiday condos. Vines hanging from trellises for shade. The sloped roof tiled in solar panels. A communal cob oven out here in the rear courtyard where the kids learn to bake. Gleeful German five-year-olds streaming out the main entrance—there was Gesine's daughter Stevie now—and onto automobile-free streets, strolling toward their ultra-efficient, mostly solar-heated homes, there to munch on exquisite *Brötchen* from the neighbourhood bakery and continue the process, evidently already under way, of evolution onto a higher plane of existence.

Sure. *Ja. Sehr gut.* This is Vauban, after all, where *Time* magazine named each of the 5,000 residents a 2009 Hero of the Environment. They earned the laurel for the simple virtue of living in gentle, enlightened harmony with the earth in this extraordinary suburb, which was erected over the past decade on the ruins of a Cold War French military base on the outskirts of the storied Black Forest city of Freiburg. And so, yes, there is much here to delight in. There is Gesine, and there is her English-bred partner Ian Harrison and their three lovely preteen girls, and there is the music Ian and Gesine make together on medieval instruments by way of a career. There is their cozy *Passivhaus*, a marvel of solar heat retention that, like every other dwelling in Vauban, requires no more than 30 percent of the usual energy budget for a German home. There's the general store around the corner, just a couple of years old but already the bustling hub of the community—"the spider in the web," Ian calls it with offhand, Vaubanesque pride—and the frequent trams gliding quietly through on tracks covered in lush grass. There is the Wagenburg squatter community, sewn seamlessly into the social fabric along with students in hyper-efficient apartments over here, seniors in ultra-accessible ones over there, and artists and designers and craftspeople in studio spaces around the way. There are the wide, ample biking and walking paths, the communal play areas, the little copses of forest primeval here and there with tree houses and swings built right in, and the chicken coop next to the district heating facility, where your kids can learn how to handle live poultry, and you can leave fifty enlightened post-national euro cents if you need an egg.

Yes, there is this whole Vauban scene, as if some Teutonic wizard had ransacked the dreams of every idealistic urban planner in the free world and stitched together all the bits and pieces of walkable, mid-rise, mixed-use, transit-friendly, eco-conscious design in the lee of a Black Forest hillside as the setting for a fairy tale called *Little Green Riding Hood Rescues Hansel & Gretel and They All Flee the Dark Forest to Live Together in Solar-Powered Social-Democratic Harmony So Luminous It Convinces the Wolf to Self-Domesticate and Form a Limited Partnership with the Witch to Provide Efficiency Retrofits at Reasonable Prices*. Yes, yes. All that. Lovely. *Wunderbar.*

But you see, Gesine told my wife and me, if we'd come at the end of last week—well, the kids were harvesting and cooking the potatoes they'd grown for themselves. And wouldn't that have been, you know, double-plus lovely? *Superwunderbar?*

I know this part of Germany. Or I thought I did. My father, a fighter pilot in the Canadian Air Force, was posted to southwestern Germany in the early 1990s—first at CFB Baden-Soellingen, the Canadian base an hour up the Autobahn from here, then a bit farther north, at the mammoth U.S. base at Ramstein, for a NATO job. I worked summers and spent my holidays in the region as an undergrad, gawking at Gothic cathedrals and drinking *Glühwein* in the Christmas markets.

Those bases were fully steeped in the folklore of the postwar
Pax Americana, thickened in those days with the giddy first flush
of fallen-Iron-Curtain victory. Freiburg was a pleasant day trip, a
museum piece full of winding cobblestone laneways from which to
peruse medieval battlements. Germany in general was a pair of man-
gled orphan twins still learning to walk again after the horrific excesses
of their deranged parents, Nazi and Communist alike. There were
variations on the theme farther afield. Spain was a charmingly underdeveloped backwater, just beginning to recover from its own escape
from the fascist boot. France was a theme park with union problems.
Italy was a theme park with management problems. Scandinavia was
a tidy but rather glum branch plant of the emerging global economy,
specializing in safe cars and disposable college apartment furniture.
And who knew how long it'd be before anything east of the Elbe
resembled modernity? These were fine old countries, sure, picturesque on the front of a postcard, but it was North America that was
going to lead the way—culturally, politically, and industrially—out
of Europe's dark century.

On the edge of the main residential area at CFB Baden-Soellingen
stood a small retail strip with a German-style café-bar above it—the
preferred hangout for Canadian teenagers. I was a regular there in
the summer of 1992, occasionally finding myself amid a table or two
of young labourers recently arrived from the collapsed GDR. They
favoured denim jackets and heavy metal T-shirts, and between our
fragmented German and their fumbling English we figured out that
they liked the place because it was so Western—by which they evidently meant non-European. North American. The next frontier.

Our conversations were inevitably disjointed and pilsner-hazed,
but what I remember is that every one of those East Germans had
vague ambitions of emigrating to America, and that when you asked
them why, they described a place that sounded like a Coca-Cola ad
randomly spliced with clips from rock videos and action movies. I
remember even more clearly the sense of offhand pride I felt at living out someone else's fantasy. They kind of wanted to *be* me. It was
intoxicating.

Not twenty years later, I stood in the courtyard of the kindergarten
in Vauban and felt something akin to vertigo at the dramatic inversion

of roles. I also felt something I'd never experienced in Germany: naked envy. Worse than that—I felt deprived. Underprivileged. Needy. If only my kids could look forward to attending a school this lavish in its amenities, this thoughtful in its design, this enlightened and *new*. I felt a little bit like some miserable wretch on the deck of an old immigrant steamer, wrapped in a tattered blanket against the maritime chill, gazing in wonder at the New York skyline.

It wasn't just the solar-tiled, potato-farmed kindergarten. I could expand my purview in concentric circles of awe: from Vauban to Freiburg, overseen by the first Green Party mayor of a major German municipality; from the municipal government to the federal, which passed the world's most ambitious renewable energy legislation in Berlin's revamped Reichstag (it now produces more energy than it consumes, thanks to a Norman Foster retrofit); and from there to western Europe as a whole. A great chain of innovation stretching from Scandinavia to the south of Spain, ultimately encompassing all the essential infrastructure of our brightest possible future.

If you accept the premise (and I do, as basically all of Europe does, despite the diversionary tactics of an intransigent chattering class here in North America that would have us believe a poorly worded email or two negates a hundred years of scientific inquiry since the greenhouse effect was first detected) that beating climate change and ending fossil fuel dependency together represent the defining challenge of the twenty-first century, then its first, tumultuous decade gives every indication that the innovations needed to overcome that challenge will happen in Europe. After a century as Western Civilization's primary battleground and museum of antiquities, Europe has again become its pace-setting think tank and laboratory.

Greetings from... The New Grand Tour

Starting in the late 1600s, a Grand Tour of European culture was an essential part of a young English aristocrat's education. The Enlightenment had come late to Britain, and the country's universities were falling into near-obsolescence. (Christ's College at Cambridge, for example, which would in time educate Charles Darwin, admitted only three freshmen in 1733.) And so the Grand Tour was under-taken in lieu of better schooling at home. A proper gentleman com-pleted his education by taking in the sights of Paris and Rome and Venice, perhaps stopping on the return journey in Vicenza to admire Palladio's villas, and in Geneva to bask in the brilliant gaze of Voltaire or Rousseau.

The emphasis, however, was primarily on Italy's ancient ruins. "The Classical heritage," writes Grand Tour historian Roger Hudson, "*was* Civilization, to all intents and purposes." Notwithstanding such earnest goals, the Grand Tour was as much about oat sowing (in legendarily debauched Venice, in particular) as edification, and for many gentlemen-in-training the ultimate goal was simply to return to England exhausted by travel and Continental peculiarities, and fully prepared to conform to the tidy norms of home—"to make them see," as one aristocratic matron put it, "that nothing is so agreeable as England." In time, the English gents were joined by upstart American boys—the White House and U.S. Capitol building were inspired by architect Benjamin Latrobe's Grand Tour study of Palladio's work—and by the mid-nineteenth century the first mass-market tourists had codified many of the Grand Tour's biases into guidebook dogma.

I'd expected from the outset that my European journey would be a different kind of tour. I'd come with my wife and two kids for the express purpose of documenting something new: the develop-ment of a sustainable industrial order. I planned to interrogate a great many technocrats on the construction of solar plants and the reconfiguration of electricity grids, while my wife anticipated wrest-ling with how to take uncommon pictures of wind turbines. We didn't count on Vauban's enviable kindergarten, but by then, around the midpoint of our trip, we'd come to realize we were witnessing something much deeper and more significant than a gallery of new and improved Continental mousetraps—something that couldn't be counted entirely in kilowatt hours or tonnes of carbon dioxide emit-ted. We'd found ourselves traversing the vanguard of Europe's Green Enlightenment on a new kind of Grand Tour.

The Rousseau of this new Tour turned out to be Hermann Scheer, co-author and global champion of Germany's renewable energy law (passed in 2000, and since copied by governments across Europe and,

most recently, in Ontario), and to my mind the single most important progressive politician alive today. "This is a structural change," he
told me, "and each structural change leads to new winners and old
losers." He was referring specifically to the change from non-renew-
able to renewable energy, predicated as it is on a shift from scarce and
expensive primary fuels like oil, coal, and uranium to free and ubiquitous fuel sources like the sun and wind; and from centralized, oli-
garchical control to a highly decentralized energy regime. But really
he could have meant the whole greenward Continental drift. This
wasn't the swapping of one energy source or even one industrial base
for another; this was the weaving of a whole new socio-economic
fabric. And the real purpose of the New Grand Tour was to gawk at
its marvellous texture and design.

Greetings from... The Most Livable City in the World

"Quality of life"—"livability," for short—is a highly subjec-
tive term. What qualities? Whose life? Measured how? In any case,
Denmark and its elegant capital, Copenhagen, frequently win international rankings of such matters. The Danes topped two recent sur-
veys of the world's happiest people, for example, and Copenhagen has
placed either first or second on the last three worldwide Most Livable
City lists published by the jet-setting British magazine *Monocle*. To
these, I would add a singular measure by which the city has revealed
itself to be an oasis of livability nonpareil: preschooler jet lag.

Copenhagen was the first port of call on our New Grand Tour,
and I arrived with my four-year-old daughter a few days ahead of my
wife and infant son, thus to squeeze in a trip to Legoland ahead of
the working part of the working holiday. To land in a foreign capital
with a four-year-old after twenty hours in transit is a bit like waking
up half-drunk behind the wheel of a moving backhoe with a jug of
nitroglycerine precariously balanced in the scoop.

Fortunately, Copenhagen's airport is about as user friendly as
they come. There are not just complimentary carts for carry-on bags,
but complimentary strollers. There is lots of soothing blond wood,
and everyone's English is better than yours. Both the metro and the

regional train network have platforms and ticket kiosks right in the terminal. And as I learned first-hand, it doesn't matter if you buy a ticket for the wrong one while you're corralling your ticking time bomb of a daughter, because it's valid on either system.

We'd rented a flat near the main train station through a Craigslist ad, and perhaps owing to inflated world's-most-livable-city standards, the place had been significantly undersold. It was bigger, better furnished, and more centrally located than advertised. The owner had mentioned in passing that there was "a playground nearby," by which she meant that the north wall of her apartment block, formed by the ramparts of the old royal shooting gallery, had been converted into the best urban jungle gym my daughter had ever seen. It had a zipline swing on fifty feet of wire, a giant sandbox, a towering rope pyramid, a wading pool. And on weekdays, a team of caregivers ran indoor and outdoor enhancement activities out of a small shed. Free of charge. With considerable flair.

The day after our arrival, we passed a long, lazy morning at the playground. The jet-lagged preschooler, however, is a capricious creature, even by the formidable standards of four-year-olds in general, and the question of how to make it through the afternoon loomed over the proceedings. I didn't want to travel far, for fear of agitating my daughter with monotony, and I didn't want to pay a Scandinavian-priced admission fee for anything one or the other of us would be too tired to see through. We were without toys or comprehensible TV. What to do? Copenhagen's livability, I knew, was predicated in part on it being one of the most pedestrian-friendly metropolises on the planet. So we simply walked. And it was *magic*.

Livability, it turns out, is a broad, car-free plaza in front of city hall, crowded on this day, serendipitously, by singing, dancing Chilean soccer fans. (My daughter was reasonably sure this was a show being staged for her benefit.) Livability is the movable feast of the Strøget, central Copenhagen's high street, which first cleared its cobblestones of automobiles in 1962, in time becoming the backbone of a network of pedestrian-only avenues and lanes billed as Europe's most extensive. Livability is a passing parade of street performers and ice cream vendors, tidy squares every few hundred metres with a fountain to climb on or a broad expanse to chase pigeons across. Livability is the temporary exhibition (yet more serendipity) set up in one of those squares, an assortment of multicoloured shipping containers retrofitted as miniature performance spaces. Livability is your four-year-old sitting for fifteen minutes in preternaturally still concentration inside one of these spaces, listening to a Danish guitar virtuoso play some enchanting baroque composition. (Free of charge. With considerable flair.) Livability is a great old ship's anchor that doubles effortlessly as a climbing gym—this mounted at the head of Nyhavn, the row of cafés housed in old candy-hued warehouses along Copenhagen's waterfront, which serves as the exclamation point at the Strøget's terminus.

Livability is a jet-lagged parent in the heart of a busy foreign city, able to relax entirely even as the preschooler darts deliriously ahead, because it is a gentle, sunny afternoon, and there are no fast-moving, thousand-pound steel boxes to watch out for. Livability, yes, is the space to effortlessly create a yawning afternoon's worth of serendipity.

"It's hardly a coincidence that the number one amendment to the American Constitution emphasizes the right to free speech and to peaceful gathering with your fellow citizens. That is one of the strongest expressions of the importance of the public space." This was Jan Gehl, professor at the Royal Danish Academy of Fine Arts and chief disseminator of the Copenhagen approach to urban design worldwide, speaking with me the following week. If Hermann Scheer is the Green Enlightenment's Rousseau, then Gehl is its genial Danish Voltaire, his consulting firm hired by cities worldwide—from Barcelona to Oslo, from Melbourne to New York—to teach them how to escape the antiquated shackles of modernist city planning and car-centred urban design and become more like Copenhagen.

Gehl believes urban public space is the lifeblood of democracy, the essence of humanism, and the *sine qua non* of green-minded livability. "Throughout history," he told me, "public space had three functions: it's been the meeting place and the marketplace and the connection space. And what has happened in most cities is that we forgot about the meeting place, we moved the market space to somewhere else, and then we filled all the streets with connection, as if connection was the number one goal in city planning, in public space." What he means is that we replaced public squares with parking lots, enclosed and privatized our marketplaces as shopping malls, and then turned over our streets almost exclusively to rapid transportation by private vehicle. In so doing, we enslaved ourselves to oil, choked ourselves on exhaust, and shattered into a million fragments the public realm where civil society once flourished.

Copenhagen's great lesson for the New Grand Tourist is that the essential first step, maybe the only critical one, in reassembling these shards and building the urban foundation of the Green Enlightenment is to put people ahead of their cars and public spaces ahead of private ones in the planning priorities of the city—of *any* city. It is so deceptively simple, so modest, that it can seem insufficient to the task. Surely something more ambitious and august is required, some monumental summit or epochal declaration. This, in any case, was the goal of the world's purported leadership when they descended on a conference hall on Copenhagen's fringe a few months after my meeting with Gehl, there to argue with awesomely ineffectual ostentation about how to avoid the collapse of civilization at the hands of climate change. They'd have done better, really, if they'd just rented bicycles and re-enacted Jan Gehl's forty-fifth wedding anniversary.

Gehl: "My wife and I, in our early seventies, we did our twenty kilometres through the city, through all of the nice places in the city— on our bikes, in leisurely tempo and a good style, on safe bicycle lanes—and had a wonderful dinner at an outdoor café. Which was one of 7,000 outdoor seats. And all the bicycle lanes and all the outdoor eating has happened while we were married. We could not have done that forty-five years ago."

My family and I followed Gehl's lead. We rented bikes from a place called Baisikeli—which naturally uses the revenue to fund an NGO that ships free bikes to urban Africa—and we strapped the kids into the dual seat on the front of the cargo bike in the pair, and we rode. We biked past royal palaces and offshore wind farms, down bustling commercial arteries, and through the forested grounds of Christiania, urban Europe's last great commune. We stopped to watch the tricks at a skate park along one stretch of newly revitalized harbourfront, and we pulled into a bike shop to contemplate buying a cargo bike of our own.

We biked mostly amid a steady flow of fellow cyclists; 37 percent of the city's residents and 55 percent of downtown dwellers travel to and from work by bike. And we rode almost exclusively in dedicated bike lanes, which as Canadians we'd come to believe consisted of a stripe of paint on the edge of a busy roadway, or even just a pictograph of a bicycle floating helplessly among parked and idling cars (serving mainly as practice targets for passing motorists). Copenhagen's bike lanes, by a comparison so stark it makes little sense to use the same term, are flaw-lessly designed and maintained, with physical barriers such as curbs, medians, and parked cars between them and the motorized traffic, and their own traffic lights at major intersections. Abandoning twenty years of habit, we conformed to local custom and rode without helmets; eventually, we didn't even bother strapping our daughter into her seat. Never once did I feel any less than the safest I'd ever felt riding a bike in a city. In the years since its main street was first closed to vehicu-lar traffic, Copenhagen has embarked on the construction of probably the best urban biking infrastructure in the world (it wrestles sportingly with Amsterdam for this crown). We navigated it with the same naked envy we encountered again and again on our New Grand Tour.

One fine Copenhagen morning, I set off on my own to meet with Mikael Colville-Andersen, proprietor of several websites docu-menting biking life in the city, and lately an ad hoc consultant to the flocks of urban planners from around the world who come to study its post-automotive model. We biked together to the broad baroque bridge where Copenhagen's busiest commuter avenue crosses a canal on its way into the city centre, and parked ourselves on a bench at one end, in front of an LED sign that counted bike traffic. There we watched the traffic and Colville-Andersen gave me a sardonic primer on the city's stylish bike culture.

"Here, the bicycle is a vehicle," he explained. "It's a tool. We have 500,000 people who ride every day, and I always say we don't have any cyclists in Copenhagen. None of them identify themselves as cyclists. They're just people who are getting around the city in the quickest way."

As with all the best design, urban design is at its most exemplary when it is mostly invisible, intuitive, inevitable. The road we were watching was the city's latest and greatest case in point. The reason they'd installed the big LED counter here was that the avenue had recently been desig-nated Copenhagen's first Green Wave artery, its traffic signals timed to provide bike commuters with a journey free of red lights. It's a commonplace technique, of course—few suburban commutes in North America do without such synched lights—but the choice of priorities is many emissions-free, self-propelled kilometres ahead.

Which reminds me of one of the most striking things about Colville-Andersen: though Danish by heritage, he's not a Copenhagener by birth. Quite the contrary; his father, like many Danes surveying a continent in ruins in the aftermath of two epic twentieth-century wars, looked west across the Atlantic for his brightest future. He settled eventually in a bustling new temple to the nascent North American car culture: the city of Calgary. Colville-Andersen grew up in an archetypal Canadian suburb just a few kilometres from my own home. All you have to do is look at the guy, though—the stylishly dishevelled hair, the hip shades, the custom cargo bike—to know you're in the company of someone who likes to be at the leading edge of the zeitgeist. No wonder he emigrated to Copenhagen.

Greetings from... The New Green Industrial Heartland

FOR MANY A SEVENTEENTH-CENTURY Grand Tourist, the reward after many weeks of hard travel and study was a sojourn in Venice. The city was by then starting to fade as a commercial and industrial powerhouse (its boatyard is the origin of the term "arsenal"), but it remained a stylish city of art and opera and epic masked balls, alongside countless other strains of cosmopolitan indulgence (the words "casino" and "ghetto" also once described singularly Venetian phenomena, and by many accounts the word "brothel" probably should have).

The New Grand Tour's closest analogue is probably Berlin. Still littered with the ruins of the Nazi and Stasi regimes, the contemporary city is a sort of life-sized sequential diagram of enlightened green urban revival. It is nearly broke but rife with ultra-modern new buildings and state-of-the-art transit. There are notable art galleries and elegant little cafés on every block, yet the rents still make it the preferred European capital for cash-strapped artists and trust-funded poseurs alike. There are Communist-era blocks of bland flats with brand new green roofs. Remnant chunks of the Wall and sombre

World War II plaques still haunt the city, but its nightlife is among the most ceaseless and exuberant anywhere. Germans are famous for inventing just the right compound word to describe something ineffable. They should make one up for the sensation of being simultaneously exhausted by the past and ecstatic about the future. In English, let's call it Berlinism.

We steeped ourselves in Berlinism at its epicentre: the funky downtown neighbourhood of Prenzlauer Berg. Another overly modest Craigslist ad had secured us another ideal apartment, though in this case the preschooler heaven of a local park was a full half-block away, in the shadow of a historic water tower that had served for a time as a makeshift concentration camp. There's Berlinism for you in a single incongruous line: the old concentration camp is really a wonderful place to bring your kids.

There might be no other urban district in the industrialized West as ravaged by the twentieth century's excesses as poor Prenzlauer Berg. Its capsule history reads like a litany of the sins of the industrial age— from prewar tenement slums to concentration camp deprivations to forty years under the boot of the East German police state. By 1989, it was so exhausted, so bereft of sustainable life, that many of its elegant old flats were abandoned, fully furnished, by East Germans fleeing west when the Wall came down. Yet today, just twenty years removed from its wholesale desertion, Prenzlauer Berg is as livable a neighbourhood as you'll find anywhere. Among many other blessings, it is now home to the best farmers' market I've ever overindulged at.

I could go on for some time about the Markt am Kollwitzplatz— the bread, the chanterelles, the olive oil pressed on site, the currywurst lunch with beer poured from a tap and served in a real glass, despite the takeout counter. To do so would really be to make a larger point about the New Grand Tour in general, thereby restating a well-known fact about Europe: that the food is, you know, weak-at-the-knees good. But moreover, that the *culture* of growing and eating food in Europe has largely skipped the most severe deprivations of modern agribusiness, meaning that even the most workaday greengrocers and mini-markets and takeouts deal mostly in what the huddled North American masses have come to think of as gourmet food. The Grand Tourists had their Venetian galas, but I found my own private decadence—even a hint of transcendence—in a sausage *mit Pommes* and a big glass of *Hefeweizen.*

An extreme example to underscore the point. There is a German pastry chain called Wiener Feinbäcker that you find in shopping malls, train stations, airport concourses. The location my family came to know was the one at the Alexanderplatz station. Like all Wiener Feinbäcker outlets, this one had a long glass counter that encased great piles of cherry Danish (which got my wife addicted) and croissants (which hooked my daughter). It also served passable coffee and, on a high shelf above the till, bread. Real bread. Big wheels of hearty German rye and sourdough, oblong baguettes and Italian loaves, plus bins filled with six varieties of *Brötchen*—Germany's ubiquitous, delectable little dinner rolls. Wiener Feinbäcker is mass-market fast food. It's the Tim Hortons of German baking. And yet its bread kicks the bland, dry stuffing out of I'd guess 98 percent of the places I've ever bought bread back home. The Alexanderplatz station has better bread than almost all of Canada. This is my point.

I wasn't in Berlin for the food, though, nor even for the general bonhomie of Berlinism. Like a Grand Tourist in Rousseau's Geneva, I'd come to soak up the wisdom of Hermann Scheer, long-serving Social Democratic Party MP and co-author of the German renewable energy law that had done more than any other piece of legislation anywhere to launch the global green economy. We met at the Café Einstein, across from the parliamentary office block on the Unter den Linden, beneath trees that shaded triumphant Nazi martial parades in the years of Scheer's infancy. He explained to me in his playful baritone growl how he'd launched the German feed-in tariff that has rapidly rewritten the energy policy of most of Europe.

(The New Grand Tour is not the occasion for policy wonkery, but briefly: Starting in 2000, the feed-in tariff, championed primarily by Scheer and Green Party MP Hans-Josef Fell and passed by the Red-Green coalition government of their two parties, set prices for electricity from renewable sources at substantially higher rates than those for power from conventional sources, and guaranteed those rates for twenty years. This not only sparked a massive nationwide boom in new wind and solar installations, but also moved the world-wide investment and production hubs of both industries to Germany, almost overnight. There is no taxation involved—the added cost is distributed to ratepayers nationwide at a price per kilowatt hour of electricity—and the estimated total cost to the average German household is €40 per year, amounting to a 3 percent surcharge.)

It began, more or less, with Scheer's vote against ratifying the Kyoto Protocol, which made him the only German MP from any party to do so. "It is not a real locomotive," he told me. "It's a barrier." Scheer recognized as soon as he learned about the gravity of climate change in the late 1980s that solving it would require a fundamental restructuring of the global energy economy. When Germany passed a weak after-thought of a feed-in tariff for wind power in the early 1990s, Scheer recognized it immediately as a potentially powerful lever.

For the rest of the decade, he slowly built alliances, helped establish a law journal to develop the legal foundations for the tariff, and squeezed a 100,000 Solar Roofs commitment through the Bundestag. With the election of the Red-Green Coalition in 1998, the drive for a muscular feed-in tariff began in earnest. It was passed in 2000, amplified in 2004, and reconfirmed by Angela Merkel's Conservatives (who had opposed its original passage) in 2008. "We have the critical mass, and we have the public support," Scheer explained. "In the renewable energy sector, there are more turnovers than in the conventional power sector now, more new investments."

The fruits of Scheer's labour crop up everywhere on the German landscape. Keen to give my kids their first real taste of quick, comfortable long-distance rail travel, we traversed the country from Berlin in the northeast to Freiburg in the far southwest by train. Wind farms and vast fields carpeted in solar panels were as much a part of the scenery as castle ruins and red-roofed villages. We skirted the rim of Solar Valley, the newly christened hub of the solar industry, just south of Berlin, where thousands of new jobs in the manufacture of solar panels have finally brought the former GDR's chronic unemployment problems under control. We passed through whole cities where passive solar design has become a part of the building code.

As we veered south at Wolfsburg, gateway to the Ruhr Valley, Germany's historic industrial heartland, I thought of my father's stories of flying over the region in the late 1960s, when it was all but permanently obscured by clouds of smog. And then I recalled recent press releases I'd read about old Ruhr coal mines covered over with vast solar arrays. We didn't make it to the North Sea coast, but German trade officials in Berlin had shown me pictures of the Seven Wonders scale of the work now occupying the city's ports: the construction of towering five-megawatt wind turbines and the 45-metre-high, 710-ton steel tripods that anchor them to the ocean floor. In the next ten years alone, Germany intends to develop 10,000 megawatts of wind power in the North Sea, a generating capacity only slightly less than the entire current capacity of my home province of Alberta.

All told, Hermann Scheer's ambitious policy innovation has converted 15 percent of Germany's electricity grid to renewable power, and created a quarter of a million jobs just since the turn of the century. One study claims that by 2020, the German clean technology sector will be bigger than the automobile industry in the land of BMW, Porsche, Mercedes, and Volkswagen. And more than that, Scheer's policy has the fervent support of hundreds of thousands of German homeowners who operate their own little rooftop power plants. You could see them from the train, particularly in the sun-kissed southwest, where the German villagescape of my memory has been given a wholesale photovoltaic overlay—a crazy carpet of glinting silicon and glass tidily laid across seemingly every other roof, in as distinct a sign of historical moment as the steeples of medieval churches.

At its very best, the feed-in tariff's transformation of the German home may be the most quietly awesome sight on the whole New Grand Tour. And probably the best place to witness it—but of course—is Vauban, Freiburg's sainted new suburb. Across the avenue from the redeveloped military base that forms Vauban proper sits Solarsiedlung, an orderly community of fifty townhouses, each of which produces more energy than it consumes over the course of a year. I'd first spoken to its architect, a visionary by the name of Rolf Disch, several years before, but I'd always wanted more intimate confirmation of the miracle of the Solarsiedlung. I was introduced on this visit to a couple who lived in one of the homes. After a half-hour of pleasantries, I asked to see their energy bills. They laid them out for me on their dining table with just a trace of bemusement.

In 2008, Harald Müller and Barbara Braun paid €398.69 (about $560) for their electricity consumption and €332.81 for their heat consumption. The same year, they were paid €3,750.29 for the electricity produced by the solar panels on their roof. Their net revenue totalled €3,018.79. They estimate that they're still a few years from fully paying off their household power plant, but by 2012 or so they'll be looking at more than a decade of pure profit.

This is how a feed-in tariff works: it turns an ordinary house into an engine that generates over four thousand Canadian dollars a year in profit. Surely there can't be a homeowner anywhere, in any circumstance, of any political persuasion, who'd stand against that. This is Hermann Scheer's confident bet—one he has made not only at home, but also in government halls willing to listen, from Spain to China to Ontario. It could indeed be as epochal as Rousseau's embrace of the social contract.

Greetings from... The Spires of Andalusia

From a certain angle, the eighteenth-century Grand Tour was simply a series of jaunts between cathedral spires. In nearly every city, the main attraction was a great Gothic edifice, compelling an obligatory tramp up the spiral steps to the cupola to admire the view. There's much less of this sort of thing on the New Grand Tour—as monolithically impressive as wind turbines are, there's no real reason other than maintenance or hubris to surmount one. The Solar Spires of Andalusia are a notable exception. Few events on the tour are as exhilarating as ascending to one of their observation decks.

Technically speaking, the spires are not spires at all. They are power plants. There are three of them—PS10, PS20, and AZ20—and they soar over a broad stretch of Andalusian plain just west of Seville. They are great concrete wedges, pale yellow or grey, with elongated hexagonal openings running most of their length (to reduce wind shear), each with a deep indentation at the top like a gaping eye. They stretch to 90 to 150 metres, or about the height of the spires on Gothic cathedrals, and each is surrounded by hundreds of giant mirrors the size of barn doors. The mirrors concentrate sunlight on the eye at the top, where a steam turbine is mounted. Ancient technology, in a sense—the Industrial Revolution began with the steam engine—but these are some of the first industrial-scale engines to be powered exclusively by the sun's rays.

The place, known as the Solùcar Platform, was erected by a sub-sidiary of Spanish energy giant Abengoa to generate 300 megawatts of electricity for the national grid and develop the next generation of solar technology. The three great spires themselves generate just fifty of those megawatts. The real workhorses are the parabolic trough concentrators: long, ditchlike assemblages of U-shaped reflective glass that concentrate light on a tube of liquid mounted at the U's focal point, thus powering a steam turbine. The troughs have been so suc-cessful that Abengoa has contracted to build another installation an order of magnitude larger in the Arizona desert.

We arrived at the platform on an unseasonably cool and cloudy September day by local standards, which meant it was merely very warm and only occasionally blindingly sunny. Owing to a last-minute babysitting snafu, we had our son in tow, and as we waited for the ele-vator to take us to the observation deck at PS10's mid-station, the guide noted cordially that ours was the first baby ever to visit the observation deck. We beamed with pride, though as it turned out he was wholly and vocally unimpressed with the amount of wind and noise.

From above as below, Solùcar was breathtaking. It plays tricks with the light that I'm reasonably sure have never existed before. When the sun ducks behind the clouds, for example, the mirrors have to be tilted away from the spire's peak so they don't reheat it too quickly when the sun returns. (Almost all the solar tower R&D at Solùcar is now focused on more heat-resilient materials.) When the sun's rays first re-emerge to strike the tilted panels, the light is concentrated at a spot several dozen metres in front of the spire's eye, at the same altitude. It appears for a long moment as if the spire has somehow extended a visible energy field before it to catch the sunlight.

Solùcar looks like a sci-fi movie set, but it also comes off as ageless and permanent, almost obvious after a while. I was reminded of a one-liner I once heard the sustainable design guru William McDonough deliver. Whenever he meets skepticism about how far we can go with this Green Enlightenment, he said, he likes to point out that it took us 5,000 years to put wheels on our luggage. It took us only a couple of hundred to employ heat-concentrating mirrors in industrial power generation. The trail ahead is thick with low-hanging fruit. And the reason all of this is happening on the plain of Andalusia is not strictly nor even primarily because it is a very hot and sunny place; rather, it is because Spain was one of the first countries to pass a conscious imitation of Germany's feed-in tariff.

My family had urgent business to attend to on a Costa del Sol beach, so I set out on my own for the final whistle stop of the New Grand Tour. I was moving along at 250 km/h in the first-class compartment of the AVE del Sol, one of the newest routes on the rapidly expanding Alta Velocidad Española (AVE) high-speed network, when a complimentary glass of dry Andalusian sherry arrived. I'd been half-reading a book, my eyes mostly on the digital screen above the compartment door that tracked our speed and progress in real time. The train topped out at 302 km/h over an unexpectedly tasty pasta lunch somewhere on the Castilian plain. I paused long enough to acknowledge that I'd never before in my life hurtled across the earth's surface at such a phenomenal rate of speed, and was struck for a moment at how I was among the first generation capable of doing so. But it was all too much to chew on after the sherry, so I leaned back to enjoy the ride.

The AVE del Sol leaves the station at noon sharp and arrives at Madrid's Puerta de Atocha station shortly after 2:30 p.m. It covers about 540 kilometres—almost exactly the distance between Toronto and Montreal—in two and a half hours. You can board ten minutes before departure, and you don't need to take off your shoes or surrender your bottle of water. Coach costs quite a bit less than most equivalent flights, first class only slightly more. The ride is smooth and sometimes truly thrilling; when two AVEs pass within a few metres of each other, going in opposite directions at 300 km/h, there is a sensation like being caught inside a muffled thunderclap. I disembarked in Madrid thinking: there has never been a combination of cost, speed, comfort, and convenience this flawless. The warm sherry buzz made the world seem luminous. My carbon footprint was 83 percent smaller than for an equivalent flight. I had just experienced the perfection of travel.

In the early 1990s, however, when construction on the first AVE line began, it was hugely controversial. The Spanish rail system was a Continental laughingstock, antiquated and legendarily inefficient, with oddly gauged trains that required pointless mid-journey transfers. Nevertheless, the Madrid–Seville AVE was seen in some quarters as little more than a fleet-footed white elephant, particularly since it was running not along the most obvious and populous route to Barcelona but straight to the prime minister's hometown. It was an age of loose spending in Spain—part of the first post-Franco upgrade in the nation's infrastructure, ahead of the Barcelona Olympics and the Seville World's Fair, and financed in large part by transfer payments from the European Union—but even still, a bullet train to provincial Seville seemed frivolous at best.

A generation later, that prime minister, Felipe González Márquez, looks like one of the world's great transportation visionaries. High-speed rail is fast emerging as *the* great global infrastructure play, and Spain is now blessed with possibly the world's best model. By 2020, 90 percent of Spaniards will reside within 50 kilometres of a high-speed rail station, and the country will be as well positioned as any in Europe to meet the transportation challenges of an unstable climate and scarce fossil fuels. (Within a year of the launch of the Madrid–Barcelona AVE line, air traffic along the route had fallen by 46 percent.) And what's more, the ease and speed of travel between major cities isn't even the most transformative thing about the AVE. No, what's truly amazing about the AVE is that it has turned the land of Don Quixote into a land of opportunity.

Until recently, Ciudad Real was a mostly forgotten regional city on the Castilian plain, 200 kilometres south of Madrid. This was the desolate landscape across which Quixote made his futile journeys (a statue of Cervantes gazes over one of Ciudad Real's main squares), and until the early 1990s the city was a place people mostly planned on leaving. That all changed when it became the first stop out of Madrid on the Seville-bound AVE line. Ciudad Real, formerly several gruelling hours from the capital by inefficient rail or highway, was now less than an hour's pleasant train ride away.

Spain's rail planners had been almost entirely focused on the big hubs, so when they noticed that AVE trains were leaving Madrid full but arriving in Seville half empty, they were initially perplexed. Further investigation revealed that a growing number of people were using the train to commute to Madrid from Ciudad Real. As the network expanded, a second category of service was introduced: AVANT, a regional network a half-step below the AVE in terms of speed (260 km/h max), aimed at commuters.

Ciudad Real soon found itself welcoming new housing developments. Its colleges and hospitals were able to attract a higher calibre of staff, since top-notch doctors and professors could now live in Madrid and commute out to the smaller city. I took the AVANT out to Ciudad Real one morning with just such a prof: José M. de Ureña, head of the planning department at the University of Castilla–La Mancha. He waved to some UCLM colleagues as we boarded. "Before, the economy was very close and very local," he told me. "Now, in a sense, Ciudad Real is becoming part of the metropolis of Madrid. And that happens a lot within one hour of commuting." Indeed, similar stories have played out along each new AVE line. Segovia: an ancient jewel over a high mountain range from Madrid, transformed from sleepy tourist town to commuter hub. Valladolid and Lleida: old industrial cities finally connected to the modern economy. Saragosa: a second-tier manufacturing city that now steals conference traffic from Madrid and Barcelona.

"The regional networks—that's the real revolution here." This was José María Coronado, a colleague of Ureña's at UCLM. Coronado was one of the school's locally sourced staffers, born and raised in Ciudad Real during its period of slow decline. "One of the biggest changes— but it is impossible to measure—is what happened in people's minds."

What he meant was that Ciudad Real saw itself now as a place in the world, a place worth investing in. Not all the commuter housing deals would pan out, and no amount of new development would transform a tired old burg on the dusty plain into a buzzing metropolis like Madrid. But then not everyone wanted to live amid such bustle. And if much of modern Ciudad Real had all the charm of a strip mall community in Kanata or Brampton (though substantially better tapas), the elegant old downtown square over which Cervantes kept eternal watch would be the showpiece of almost any city in North America. And there was now a reason, for the first time in generations, to plan to stay.

And more than this—much more than this—there was now a sense of momentum in Ciudad Real. In Spain. Across Europe. A renewed sense of optimism. Three hundred years of migration west across the Atlantic had been predicated on the belief that the empty continent on the other side was a land of infinite opportunity. The overarching lesson of the New Grand Tour, it seems, is that the centre of gravity is shifting rapidly back to the east side of the pond.

ON TIPPING IN CUBA

The Walrus, April 2012

Cuba is a nation unlike any other on earth—a distinct and ferociously proud culture, a governing dictatorship more than a half century old and totally distinct from all others of its ilk, a place stuck geopolitically in a singular wedge between the two great powers of the Cold War. And yet from the right angle on a certain kind of day, it also feels like a microcosm of everywhere else, a thin sliver of island on the global map in which all the frictions and contradictions of the global economy have been etched in high relief.

This is another story that emerged by chasing a hunch. My wife and I had visited Santiago on a brief daytrip during our honeymoon in eastern Cuba. I finagled a return trip from WestJet's in-flight magazine, wrote a standard-issue travel piece, and again convinced The Walrus *to let me ruminate at length on a stray holiday thought.*

MORE THAN A MILLION Canadians will travel to Cuba this year. The only places beyond our borders that attract more of us are the United States and Mexico. There is no other tourist destination on earth where Canadians are so dominant, and possibly none where the tourist economy is more vital to the nation's immediate economic health. With little in the way of formal policy and with no real intent on the part of the beach-bound hordes, we've established a relationship with Cuba that is unique in both our histories. We've colonized Cuba on vacation by accident.

This is a story about what happens when the unarticulated, half-hidden nature of that colonial relationship is suddenly exposed. It's an economics lesson in the form of a parable, a traveller's tale about the strange connection between master and servant in this de facto tourist colony.

So let's begin, in fairy-tale fashion, in a tower atop a castle: the rooftop terrace of Hotel Casa Granda in Santiago de Cuba, the country's second-largest city. The Casa Granda is an old colonial half ruin overlooking a wide square and an elegant cathedral. It's an atmospheric, Graham Greene kind of place, five storeys tall and colonnaded and shedding white paint. I found myself there at sunset one January evening, sipping a mojito and pondering the real value of ten convertible Cuban pesos.

Because Cuba is among the few nations on earth with two official currencies, a Neverland economy caught in its own distended bubble halfway between the collapsed Soviet bloc and the contemporary global capitalist order, visitors can find themselves wondering more than usual about exchange rates. There is the regular, nonconvertible peso, officially the Cuban peso or CUP, used to buy staple goods at state-run shops. And there is the convertible peso, the CUC— the hard currency, which is used for luxury goods and provides the default banknotes for the tourist economy. In government accounting, CUCs and CUPs are valued one to one, but informally the CUC is worth about the same as the Canadian dollar, while the CUP has a street value of a nickel at most. CUPs are worthless outside Cuba, except as souvenirs.

Earlier in the day, I'd had ten CUCs snatched from my hand, and I was up on the roof of the Casa Granda trying to figure out what exactly had happened and how I really felt about it. It's rare, once you're well into the mortgage-and-kids phase of adulthood, to encounter a whole new category of emotion, but I was pretty sure I'd done just that out there on a dusty Santiago back street, and now I was probing the feeling to discern its dimensions.

What happened, in brief, is that my wife and I had hired a young man named Antonio to give us a tour.[1] We'd spent the morning chugging around in an ancient Moskvich sedan, with another young man driving as Antonio pointed out the sights and delivered a running commentary about what he called "the reality of Cuba." We'd visited an Afro-Cuban cultural museum, toured the old Spanish fort, bought contraband rum. We'd gone back to Antonio's tiny concrete box of a home, met his wife and mother, sipped beers, talked some politics, and taken pictures. In the late afternoon, he and another friend had led us to a lovely little restaurant at the base of Santiago's landmark Padre Pico steps. We'd eaten grilled lobster, drunk more beers, and traded jokes and vows of eternal friendship.

At the end of the meal, I'd given the waiter CUC $80 and received CUC $10 in change, and as I stood there with the ten-peso note in my hand Antonio grabbed and pocketed it. I shot him a confused look, and he responded with a half shrug that seemed calibrated somewhere between *What's it matter?* and *You know the score.* I hadn't intended to give him the money, but he decided he deserved it. Hours later, on the rooftop patio of the Casa Granda, drinking a mojito that

---

1 To protect individuals' identities, some names have been changed.

cost nearly half the amount I was obsessing about, I wondered what that shrug really meant.

THIS HAD ALL OCCURRED in that informal, sketchily delineated commercial zone that springs up in pretty much any robust tourist economy, a grey market that is particularly broad and heavily trafficked in countries where visitors and locals are separated by wide gulfs of wealth, power, or political freedom. In Cuba, Antonio and I stood on opposite sides of a divide created by a substantial admixture of all three.

We were under no illusions, my wife and I. We understood that Antonio was, in local parlance, a *jinetero*—a tout or fixer, though in Cuba the word often suggests a darker meaning: hustler or scam artist. We'd participated in a drawn-out haggling courtship with him across a couple of days, a dance familiar to us from other trips to countries with bustling grey markets. There had been repeated encounters on a busy street in front of a tiny, decrepit photo studio that may or may not have been his usual place of work. He'd presented us with a gift of a photo and some cheap cigars, and we'd discussed the possibility of a city tour as if it were a friendly outing and not a paid transaction. We'd checked with the tour desk of the Meliá hotel, the only full-service resort in town, and knew that an organized tour in an air-conditioned bus would run us more than $100. We preferred the idea of giving our hard-currency CUCs and maybe a gift or two to an enterprising *jinetero* instead of to a government-controlled joint venture.

It was the right choice, and we had no regrets. It had been a fun day, revelatory in ways an official tour never could have been, and we'd been generous with Antonio and his family, paying out his fee in bits and pieces, in overpayments and unasked-for change. I'd overpaid for drinks at the Afro-Cuban cultural centre, handed him too much money by a factor of at least five to pick up a six-pack of beer to share back at his house. We'd stopped back at our hotel at the end of the tour and filled a bag for him and his family with markers, notebooks, two toothbrushes, some new towels, and a used pair of Adidas shorts. We'd paid him and the driver CUC $20 each for their work as guides—the equivalent of a month's wages in a typical government job—and had bought him and his other friend lobster dinners (another month's wages, although the place never would've served Cuban diners on their own, even if they'd had the CUCs).

So, yes: we all knew roughly what the score was. But to Antonio's mind, I'd come up with a figure at least CUC $10 too low, and he'd taken it upon himself to round it up. At first, standing in the street outside the restaurant as he and his friend departed in a flurry of hugs and warm handshakes, I'd felt something verging on a sense of violation. Had I been robbed? Had I been—that most dreaded of veteran traveller fears—*played* somehow? Was I a sucker?

With the clarity of a second mojito and the brilliant sunset over Santiago Bay, I knew that wasn't it. This wasn't a robbery or a con. This is why it felt so weird: it wasn't about what had happened to me; it was about *who I was* in Cuba. This had been a refusal to hew to the script of colonial power. It was a servant's insubordination before his colonial master. It was, I came to think, a profoundly Cuban way to read the situation.

Of course I could spare the ten pesos, and Antonio had earned it and then some. I'd seen just enough to imagine how hard it would be for him to get his hands on another CUC $10 note, and he understood all too well how effortless it was for me to obtain a big stack of them. He wasn't fully invested in the tourist economy's cold logic of wealth and status, and he certainly didn't feel he owed me any deference. He'd simply taken what was rightfully his, an inverted colonial tax on the day's transaction. Quite literally, it was what the market would bear on that particular day in Santiago.

By the time I'd reached the mashed lime and mint leaves at the bottom of my second mojito, I saw the whole thing much more clearly: me and Antonio, Cuba and Canada, the whole trip. The real exchange rate on ten convertible pesos preoccupied me for the rest of our visit and cast new light on the month's preparation leading up to it. Ten pesos was a bargain, really, for all that had been revealed.

A FEW DAYS BEFORE we departed for Cuba, we stopped by the Walmart in Antigonish, Nova Scotia, where we were visiting my parents for Christmas. We had pictures to get printed and some final purchases to make in preparation for the trip: sunscreen, medication, a high-capacity memory stick. One of the unspoken assumptions in Canada's colonial relationship with Cuba is that tourists import stockpiles of consumer goods to hand out to locals. In Cuba, unlike the rest of the beach holiday Caribbean, it's not just that people can't afford as much as we can, but that much of what we take for granted is completely unavailable in Cuba's truncated marketplace. I'd been reading on the internet about what people suggested to bring: dental floss, shampoo, good towels, baseballs, vitamins and medicines, toothbrushes and school supplies, hardware and reading glasses.

We wandered Walmart's over-lit aisles, past overflowing shelves, filling a cart with pens and notepads and econo-sized bottles of acetaminophen. We had our five-year-old daughter in reluctant tow, and she settled into a shelf piled with towels as if it were a bunk bed while we had an absurd debate about which of them to buy. The towel aisle had at least four distinct gradations of price and quality. I'd written "high-quality towels" on my list and put a big asterisk next to it, but I couldn't remember whether I was reminding myself that towels were so useful we should buy as many as possible or that Cubans needed particularly high-quality towels. In the end, we bought two bath towels ($5 each), four washcloths ($2 each), and two hand towels ($4 each). By the time we reached the checkout, I could no longer remember whether they were higher in the quantity or the quality range. The total bill was $186.82, the equivalent of nine months' average wages in Cuba. We paid with the swipe of a card and a tired shrug.

This, of course, is the way of the Canadian economy nowadays. We're delirious with choice, so completely buried in our own abundance that it inspires reality TV series. Many of us rarely bother with cash; some digital chirrup races at the speed of light from Walmart's till to a server farm representing the bank's vault, and by the time we hit the automatic exit we can't remember whether we paid $176 or $186 for the dead weight filling our cart. You could lose CUC $10 just by forgetting to put back that pack of fancy pencil crayons your kid tossed in while you were distracted.

For all the arbitrary afterthoughts that govern its periphery, Canada's relationship with Cuba, economic and otherwise, is a significant one. Canada and Mexico were the only Western nations that didn't cease diplomatic relations with Cuba during the tense missile crisis years of the early '60s, and Fidel Castro was close enough to Pierre Trudeau that he served as an honorary pallbearer at the former prime minister's funeral in Montreal. In the annals of Canadian diplomacy, there is no other international relationship in which Canada has stood as far apart from the U.S.

As a reward for our enduring friendship, Canada is probably the second most important economic ally Cuba has after Venezuela, which supplies more than 60 percent of the island's oil. (China exports more stuff to Cuba, but the Chinese don't show up daily by the multiple charter-flight-loads to hand out gifts and pour hundreds of millions of dollars into the Cuban economy.) Cuba is our largest trading partner in the whole of Central America and the Caribbean. We export a range of commodities to Cuba—sulphur, wheat, copper wire—and we are the second-largest buyer of Cuban exports, particularly sugar, nickel, fish, citrus fruits, and tobacco. That's over $1 billion in total trade. The Canadian mining company Sherritt International has a huge presence in Cuba, digging up nickel; and Toronto-based Pizza Nova, the only foreign pizza joint in the country, has operated six outlets scattered across the island. Back home, meanwhile, souvenir shops in many Canadian cities feature humidors well stocked with Cuban tobacco products for sale to American tourists. The Cuban cigar has become, in a sense, as Canadian as maple syrup.

The economic view from the other end of this relationship is nowhere near as rosy. Since the collapse of the Soviet Union in the late '80s, Cuba has struggled through a punishing "Special Period" during which its economy has existed in its own isolated socialist limbo: tethered to a failed Communist order, still barred from regular trade with the world's largest economy just 145 kilometres across the sea, desperate for imports and the hard currency to pay for them. As Soviet goods vanished from Cuban pantries in the first years of the Special Period, the average Cuban's calorie intake plummeted by 30 percent, and the local diet has never fully recovered. Cuba remains reliant on imported goods (increasingly, under a special exemption, from the U.S.) for somewhere between 60 and 80 percent of its food. Cubans also switched almost overnight to organic and in some cases pre-industrial forms of local food production; the island has experienced a boom in the production of yokes and plows for use with oxen, for example, among many other austerity measures adopted to weather the Special Period's rough, uncertain seas.

Austerity's limits, though, are now lapping perilously high against the makeshift vessel's sides. Since June 2009, Cuba has operated under emergency energy quotas in a program known as "Save or Die," while the government has been laying off workers, encouraging small business development, and sending so many doctors abroad (as trade in kind for vital commodities such as oil) that there are now nearly twice as many Cubans caring for the sick overseas as there are working on the island. The government has begun to discuss phasing out subsidies for many staple goods, suggesting that the days are numbered for the ration stores that provide Cubans with much of their daily bread (and rice and beans and milk).

One recent study described Cuba's current approach as "survival economics." The Havana-based dissident blogger Yoani Sanchez, writing for the *Huffington Post* last August, summed up Cuba's conundrum more eloquently: "We are in transition, something seems to be on the verge of being irreparably broken on this island, but we don't realize it, sunk in the day-to-day and its problems... we are leaving behind something that seemed to us, at times, eternal."

In the face of this deepening uncertainty, there remains one bright and intensifying light on Cuba's troubled horizon, one safe port amid the Special Period storm: tourism. In 1990, at the dawn of this strange age, Cuba attracted 340,000 tourists; in 2011, it welcomed some 2.5 million. Since the announcement of the Save or Die program, Raúl Castro's government has introduced ninety-nine-year leases for foreign investors, to encourage resort and golf course development; and it has loosened the restrictions on family-run, home-based restaurants (*paladares*) and guest houses (*casas particulares*), to provide more self-employment opportunities in the tourist sector. Meanwhile, one in four of the island's 80,000 government tourist workers has a post-secondary degree. "Right now," writes Sanchez, "the main incentive for those who work in snack bars, restaurants and hotels lies in the possibility of a visiting foreigner leaving them some material gratification." More than a million of those foreigners, 44 percent of the total, are Canadians. The next-largest share, approximately 175,000, belongs to the British. Tipping, once regarded as counter-revolutionary and beneath the dignity of Cuban patriots, has become their most direct and vital connection to economic stability.

TO BE A CANADIAN TOURIST in Cuba is to be something more than a visitor, more even than a run-of-the-mill mark. It's not just that you're visibly foreign and rich; you're a sort of modern vassal, the only readily accessible emissary of a metropole that has never been seen but is generally understood to be bounteous and benevolent.

Such were the macro-socioeconomic forces at work as we strolled down a broad downtown *avenida* on our first full day in Santiago. The traffic on the road was steady and loud, heavy with ancient GM trucks, diesel-belching Chinese buses, and antique Fords and Cadillacs with multiply rebuilt engines growling under their hoods. We stopped to admire the stunning facade of the Hotel Rex, a glorious art deco and neon relic from Cuba's swinging '50s. The Cuban Revolution was born in Santiago, and this was the hotel where the first *Fidelistas* stayed the night before their failed 1953 raid on the nearby Moncada Barracks. (Following the attack, Fidel, forced from Cuba for leading it, would meet Che Guevara during his Mexican exile, returning a few years later and capturing Santiago in late 1958—the first major victory of the revolutionary war.)

Later that morning, a young man waved to us from a concrete stoop half a storey above street level. He'd noticed my wife's camera and invited us in to see his little photo studio. In his solid but thickly accented English, he introduced himself as Antonio.

The studio was a study in contemporary Cuban improvisation. It was located in the cramped front room of the old house, its walls crumbling and peeling in that signature Special Period way. The only light came from a single bare fluorescent tube mounted horizontally on one wall, its unearthly glare illuminating a threadbare white curtain. The sole piece of photographic equipment was a point-and-shoot digital camera that looked at least five years old and produced pictures several megapixels smaller than the ones taken by a basic smart phone. In Santiago, though, access to any digital camera evidently provided a sufficient foundation for a photography business.

Antonio and a couple of colleagues were taking pictures of a baby girl and her proud parents. Her first birthday party was this coming Saturday, and we were promptly invited to attend. We posed for pictures with the birthday girl and her family, and then they left and Antonio launched into a protracted monologue about the local photo business, Santiago ("the most Caribbean city in Cuba"), and his own Afro-Cuban heritage. He dug out his passport and showed us the visa from a trip he'd taken to Amsterdam a few years earlier. He had a couple of pictures of himself in a toque and a heavy coat in the Dutch winter. He knew a musician, he said, who had toured Canada. His ache for escape was palpable.

We left Antonio with a vague promise to return, perhaps for a tour of "the real Cuba," perhaps for the birthday party. We were under no obligation, and we knew enough to be wary, but we meant it all the same.

It was a glorious day, warm under a gentle sun, and we strolled lazily down a nearby market street, stopping to gawk at the strange array of goods in the government shops. Cheap Chinese shoes and plastic toys were displayed on shelves or under glass in half-empty cases. There was a store full of knock-off electronics and tiny washing machines. All of it was bathed in late-afternoon shadow, the lights kept Save or Die low. Locals clutching creased ration books lined up out the door at an egg dispensary.

We came upon a nook where a young woman was selling ice cream bars. A small sign read "$3.00." I asked for one and handed her three CUC $1 coins, realizing even as I dropped them in her hand that the sign was surely referring to nonconvertible CUPs. She whisked the coins away into a cubbyhole below the counter, trying hard to seem merely efficient. She handed me an ice cream on a stick and fixed me with a blank stare. I held her gaze for a yawning moment, not sure what I was waiting for. We were colonizer and colonized, deep in the fuzzy grey market of the tourist economy, wondering who would blink first. "*Gracias*," I said finally, and we walked on down the street.

A few blocks farther along, I came to understand just how much I'd overpaid. The ice cream had been chalky and flavourless and we were still peckish, so we stopped at a corner where an old woman was selling roasted peanuts wrapped in paper cones. I gestured for one and handed her a twenty-five-centavo coin, a twelfth of what I'd paid for the ice cream. She was aghast. An old man standing next to her made an exaggerated goggle-eyed gesture of shock and then scooped up as many of the cones as his two hands could hold, nearly everything on her tray. Eventually, we settled on two cones for the quarter, for which she thanked us extravagantly.

I'd handed the girl at the ice cream stall a windfall. She sold lousy ice cream to locals from a stall on the state store shopping street in a largely tourist-free city. (Unlike Havana, which draws hordes of visitors on its own and many more on day trips from nearby Varadero's crowded resort strip, Santiago is too remote for the package tour masses, three hours by road from the much smaller cluster of beach developments near Holguin.) She had no expectation of returning home that night with CUCs in her pocket—a week's wage in a single, accidental hard-currency tip. But we were, after all, Canadians on the sunny end of thousand-dollar flights, and what did a couple of misspent CUCs matter here or there? The price of almost everything was arbitrary in Cuba. Normal rules didn't apply. That was part of the fun.

We spent a week in Santiago, and only twice did we encounter anything like a posted exchange rate, both times at live music venues near the main square. The first was at the government-run Artex music store, which sells an exhaustive range of Cuban music CDs (after rum and cigars, probably Cuba's most popular souvenir and its most widely adored cultural export). A small courtyard out the back features live music all afternoon and evening, and a sign at the top of the stairs leading down to the patio lists the admission prices in both convertible and nonconvertible pesos: "CUC $1.00/CUP $20.00." Up the street at Casa de la Trova, the most storied music hall in Santiago if not all of Cuba, locals were paying CUP $25 for an admission ticket marked "CUC $1.00."

Once, walking past a ration store, I saw rice listed on the chalkboard above the long, weathered wooden counter at $0.25 per pound (CUP, of course). At Patio de Artex's twenty-to-one exchange rate, I'd paid CUP $60—enough to buy 240 pounds of rice—for one barely edible ice cream bar. We could have bought 160 pounds of rice for what we spent to watch an hour of music one fine afternoon on the patio.

Let's talk about that Patio de Artex show. It points to the most obvious reason, beyond the availability and convenience of cheap flights or the historical friendship or the incomparable quality of a real hand-rolled Cohiba, that more than a million Canadians visit Cuba every year (compared, for example, with the 754,000 who travel to the Dominican Republic, where prices are lower, the cigars and rum are equally plentiful, and the logistics are uncomplicated by Communist bureaucracy and the grim exigencies of the Special Period). It's related to why it's no accident that Cuba alone has managed to pass half a century in open, hostile defiance of the world's most powerful nation just to the north. Every country has its character and customs and quirks, but there's a depth of soul to Cuba that puts it in a class by itself. Cuban music has set the tone and the rhythm for much of Latin American music for generations, and it's the most visceral manifestation of the island's indomitable spirit.

So, yes, let's talk about the Patio de Artex show. At the precipitous cliff's edge of the Special Period, even after many grinding months of Save or Die austerity, you can still find a table at Patio de Artex on any old Friday afternoon and watch eight guys in donated T-shirts and Chinese jeans transform a courtyard into one of the best places on earth from which to launch a weekend. Mojitos sweating through plastic cups on the table, the trumpet player muting his horn with his hand to add a vampish growl to the *son* they're tearing up, a propulsion in the rhythm that hauls even a hopeless, doubly left-footed non-dancer like me to his feet—this is what you get for your 160-pounds-of-rice admission at Patio de Artex. You get escape, transcendence. The show would have been a bargain at CUC $10 a head.

This is why Canadians come back again and again. And why, perhaps, they bring even more T-shirts and towels and acetaminophen the next time: because these people deserve more for their labour. They deserve better. Yes, this is true of any picturesque beach destination in the impoverished tropics, but in Cuba it's somehow more undeniable. Maybe it's the grinding workaday cruelty of the Special Period, the utter absurdity of America's ongoing embargo. Maybe we delude ourselves, in Mexico or Jamaica, with the notion that the society's nominal freedom means no absolute barrier exists between our decadent days by the pool and the women slaving away at the messes in our hotel rooms. Anyway, there's something about Cuba that brings the arbitrary nature of wealth and power and material comfort into especially high relief. And so we bring stuff. Gifts. Offerings. Talismans of apology and absolution.

Live music is not hard to find in Santiago de Cuba. The city prides itself on being the place where African rhythms first mated with Spanish harmonies and instrumentation to produce the itinerant nineteenth-century musical tradition called *trova*, the wellspring of Cuba's world-conquering musical culture.

One evening at Casa de la Trova, we caught a fantastic seven-piece group called Ecos del Tivoli who performed in matching suits. At other shows, we heard "Chan Chan," the Buena Vista Social Club's signature tune, played about a dozen different ways. We went to the *paladar* closest to our hotel early one night, and they opened up just for us, and before they even took our orders a kindly older man in a fedora showed up with a guitar to serenade us (his yearning take on "House of the Rising Sun" was a highlight). We bought one of his *son* band's CDs (CUC $10, counting a tip for the serenade) and then saw him in the crowd a couple of nights later at Casa de la Trova, where he cajoled a prostitute into showing me a couple of moves on the dance floor.

One afternoon, as we wandered the back streets around the original Bacardi rum distillery, an old woman approached us on the corner. She explained that she worked in economics and computers but her lifelong passion was opera, and then she asked if we'd like to hear a song. As locals strolled past with little more than a glance—just another afternoon in Santiago—she treated us to a passionate mambo in a voice that revealed years of classical training.

On Calle Heredia, a side street near Santiago's main square, there is a small tourist market. Sidewalk vendors hawk handicrafts, and behind them small shops and kiosks trade in antiques, art, and souvenirs. I repeatedly visited one in particular, a place the size of a walk-in closet stuffed to the rafters with books, postcards, and old Cuban LPs. As soon as I expressed an interest in the records, the shop's proprietor, a genial senior citizen in a newsboy cap, grinned at me around his cigar stub and started yanking out records and throwing them on his phonograph. He tangoed around his shop, pouring out coffees for both of us. I bought priceless *son* records for a CUC or two apiece, and a small stack of '70s back issues of *Bohemia*—once one of Latin America's most important magazines—for the same unit price. For reasons I never fully understood, the old gent threw in a freebie, a fold-out postcard packet dated 1958, featuring photos of the La Seo Cathedral in Saragossa, Spain.

What would you consider a fair price for admission to one of the best musical theme parks on earth? Twenty bucks? Fifty? What do they charge to get into the Grand Ole Opry in Nashville? If they'd asked me, as we boarded our flight home, for CUC $10 for the old lady's street corner bolero, I'd have paid it gladly. What I felt most acutely, weeks later, is that I'd failed repeatedly to leave an adequate tip.

LET'S TALK ABOUT where to eat in Santiago (and how to under-tip there). The state-run restaurants—the CUC joints for tourists, as well as the CUP places for locals—are a complete waste of time. The food in the family-run restaurants, the *paladares*, is in another league entirely. *Paladares* were black market operations, just a couple of tables in someone's family kitchen, until the mid-'90s, when some of them were licensed and subjected to byzantine regulations and steep fees. The unlicensed ones, though, remain the best, and we found one around the corner from another *trova* hall, the Casa de las Tradiciones, simply by asking the guy at the door if he knew of a place nearby to eat.

He led us briskly up a side street, where seemingly every twenty-something male in the neighbourhood was gathered around a boisterous roadside domino game. As we approached, a young man stepped away from the crowd and introduced himself as Luis. He led us a couple of blocks farther along to a tiny bungalow and invited us to take a seat in the living room, where a woman who appeared to be his grandmother was watching telenovelas. Luis disappeared into a back room for what felt like half an hour. When he finally returned, he ushered us past the kitchen, through a narrow bedroom with kids' bunk beds against one wall, and down a concrete staircase to a small patio. A single table had been set immaculately before a panoramic view of Santiago Bay. *Trova* played on a portable stereo while our host, with a veteran tour guide's ease, rattled off historical details about Santiago's port and its slave trading history.

We noticed a small shrine on the corner of the patio, a wooden box standing on its end, the open side facing outward, with dolls, a bowl of coins, and an egg on a tray arranged inside. There was a cigar and a small wooden cross on top, and a chalk circle drawn on the concrete in front of it, around markings of arrows and skulls. Luis explained that he practised a syncretic faith called *Palo Monte*, a common Afro-Cuban religion similar to Santeria. We mentioned that it was the only shrine we'd seen. Everyone has them, he replied, but he didn't bother hiding his. He gave us a business card with the name and address of the *paladar*, the only one of those we saw as well. Amid Cuba's current evolution, Luis was more concerned with positioning his modest business for the next phase than with keeping it from the authorities in the last days of this one.

"In Cuba," he told us, "today is today. *A hoy es hoy. Mañana* is another time."

He served us a variation on the mojito, using basil leaves instead of mint. He called it an *alto del mar*—"above the sea," also the name of his *paladar*—and it was the best drink I had in Cuba. Dinner was whole fried fish garnished with the only red pepper we saw in Santiago, and a delicate creole sauce that was several steps above the licensed *paladares'* offerings in its refinement. When I asked for the bill, he brought me a scrap of paper with "$14.00" written on it. I gave him CUC $20, under-tipping for one of the most memorable meals I've eaten anywhere.

Luis's place was just a couple of blocks from the Museum of the Clandestine Struggle, which we visited a few days later. It was a shrine of another sort, its glass cases containing carefully preserved blood-stained suits and Molotov cocktails from the guerilla street war fought block by block in Santiago for years ahead of Fidel's return from exile, testimony to the long-standing Santigueran tradition of defiance.

Outside, we stopped to watch three teenagers playing an impromptu stickball game in the street. They had a broom handle for a bat, and they were hammering some kind of makeshift ball off the surrounding buildings with such authority it took a while to realize it wasn't even round. It was an empty pill bottle. I had a sudden, sick flashback to our visit to the Walmart in Antigonish and the "baseballs" entry on my shopping list. I'd searched the sporting goods section for baseballs, but in December in Nova Scotia there weren't any. I'd had a sleeve of tennis balls in my hand at one point but had put them back. With a staggering lack of perspective, guided by some deranged sense of propriety, I'd thought to myself that Cubans were world-class ballplayers who surely honed their batting skills using a properly weighted baseball. Now, though, it was the grandest of my under-tips: the one not given.

THE BRISAS SIERRA MAR is a slightly dishevelled three-star resort sixty-five kilometres down the coast from Santiago, a drive that takes nearly three hours along the craziest lunar-landscaped road I've ever encountered. The hotel is a standard-issue all-inclusive: a pool with a swim-up bar, a few restaurants and lounges, an open-air amphitheatre for nightly cultural shows, a dive shop, a beach that was really something before Hurricane Dennis reduced it to a thin strip of sand in 2005. The clientele was at least three-quarters Canadian when we arrived, a significant number of them repeat visitors who came yearly or even more often, and they lent the place a friendly summer camp vibe. The package tour operators near the front desk all kept binders full of details on excursions, and every one of them described a Cuba beyond the gates where dangerous con artists lurked on every corner; the binders all warned against any unaccompanied travel whatsoever outside the resort.

We had come to Brisas Sierra Mar to dive. The divemaster was a soft-spoken, sharp-witted family man named Edgar. We were the only divers who'd never visited the resort before, and everyone else on our excursion asked after his family. His daughters had been sick, he said, with a pointed shrug that suggested illness could go further south in Cuba than you really wanted to talk about.

Cuba's coral reefs are currently rife with lionfish, stunning creatures with brilliant black and orange stripes and a broad mane of poisonous antennae protruding from their fins. Native to the South Pacific, they are an invasive species in the Caribbean that feasts on defenceless hatchlings up and down the reefs. Edgar brought a speargun on most dives to take out as many as he could, and so one of the great spectator sports was watching the divemaster hunt lionfish. One day, he filleted a few of them back at the dive shop, expertly slicing away the venom-tipped fins using a pair of spears as an Asian cook wields chopsticks. The chef at the beachfront restaurant breaded and deep-fried the fillets for us and served them with fries: lionfish and chips, easily the most memorable meal at the Brisas during our stay.

The morning we left, I headed down to the dive shop with a plastic shopping bag stuffed with gifts for Edgar: a big bath towel, a binder and a children's notebook with a tiger on the cover that our daughter had picked out, plus bottles of ibuprofen, Dramamine, and a children's anti-nausea medication called Nauzene. Edgar had called in sick that morning, but the caretaker at the shop said I could leave the bag in his office. I went in, set it down under Edgar's desk, and then hesitated: what if someone else took it? There was no record of it, of course, no way to guarantee Edgar and his sick daughters would get it. I was invested, I realized, in the arbitrary personal connection. I wanted Edgar to know I was looking out for him. This is often the way of gift giving in Cuba—it's not enough to be the colonial master of the exotic treasures, dispensing them with a sovereign's whim. Even in the act of charity, we want tribute. We want to take a little gratitude home with us.

I left the bag in Edgar's office, knowing that whoever wound up with its contents, their value would be immediately understood and cherished. The stuff would be used and reused, used up, exhausted of its value, and then repurposed. An empty econo-sized ibuprofen bottle would, after all, make a serviceable baseball.

I wasn't ready for the condition of Antonio's home. There are more flattering ways to say it, but at the end of the tour in Santiago when he invited us back, it was so much worse than I expected. I was imagining something like the places that lined the streets on our meanderings around the city: tidy concrete bungalows, cramped but homey spaces like Luis's place.

Antonio led us down an alleyway in a densely populated neighbourhood near the photo studio. There were jumbles of concrete piled atop one another in behind the homes you could see from the street, places filigreed with rusted rebar and roofed in salvaged tin.

Antonio lived with his young wife in one of them, which was stacked on top of the slightly larger concrete slab where his mother and sister lived. The entry was up a rickety flight of repurposed wooden stairs. There were two rooms: a front living area with a sink in one corner piled with dirty dishes, and a larger back room with a mattress on the floor. There was a Chinese boom box in the front room with one of those outsize, blinking blue-green displays, and a poster of a Dutch league soccer team tacked to the wall above it. It was a makeshift place, a slum dwelling.

Antonio had had his friend drive us all over town in an ancient Soviet-made automobile mostly held together by force of will. We'd passed grand colonial mansions—"the houses of the people in Miami," Antonio explained—that now housed educational facilities and cultural centres. He'd taken us to an Afro-Cuban cultural centre and museum that we'd never have found otherwise. On the way to the Spanish fort outside town, we'd pulled off the road and waited while he ran to the stoop of a bungalow to procure two bottles of rum with legendary Matusalem labels. No one has made rum under the name of Matusalem in Santiago since Fidel nationalized the distillery, but it was premium stuff for the discount price of CUC $10 per bottle. He'd known just the right restaurant, an unsanctioned *paladar* that turned out delectable grilled lobster. Throughout, Antonio was personable, forthcoming, full of information and generous with it. He was bilingual, literate, and quick witted. His needs were obvious, unmistakable.

I forgot all that. That he was achingly poor, and I impossibly rich. That his madly uncertain future was riding on how many CUCs he had when the Special Period's precarious economic dance stopped and rendered them as worthless as Ostmarks. In the instant he took the ten-peso note from my hand, I was simply a colonial officer standing before him, momentarily indignant. There had been gifts, and there would be more, and how dare he presume to decide when and where and how they were to be bestowed?

That's the real value of ten convertible pesos: under the right circumstances, it will show you exactly who you are in Cuba. You might not like what you see, but it's still well worth the price.

The day before we left Cuba, we took a taxi to Gibara. Before the Spanish colonists built the country's first railway and placed its terminus in Santiago, the smaller city had been a major port for the sugar trade. Nowadays, it is a sleepy warren of dilapidated colonial mansions fronting on a gorgeous, breezy stretch of Caribbean Sea. While it's less than an hour's drive from the international airport in Holguin—closer, actually, than the strip of all-inclusive resorts farther down the coast—it sees only the occasional handful of day trippers on excursions. The locals are friendly, the accommodations at private guest houses in those ancient mansions charming if not quite grand, the seafood abundant and delicious. It's a breezy, laid-back, disintegrating dream of a place. Gibara is, in other words, the quintessential off-the-beaten-track hideaway, the sort of place Lonely Planet built a guidebook empire discovering.

We adored Gibara. For CUC $10 each per night, we had a comfortable room in a high-ceilinged colonial townhouse with a broad, quiet, hammock-strung courtyard in back. Within a couple of hours of our arrival, we'd already begun speculating on how cheap and easy it would be to hop a winter charter to Holguin and spend a month there living on next to nothing.

At sunset, we wandered down to the harbour to watch a ferry arrive. It disgorged a steady stream of locals in work clothes on foot, and I'd just begun to wonder if they were employees at the resorts up the coast when my wife, who was taking pictures, slipped off the edge of the curb and twisted her ankle. A crowd soon gathered around us, the faces charged with real concern. Everyone said we should go to the hospital, and someone offered to find us a ride. My wife insisted she was fine; she just needed to rest a minute.

"You're hurt," someone said. "Look at your ankle. Why not go?"

"We're Canadian," my wife answered.

"It's for everyone. Just go."

Cuba has one of the best health care systems in the developing world. Its medical schools train doctors from across Latin America and beyond, and it sends its skilled health professionals all over the globe. Cubans are, as we learned in an instant, offhandedly but beamingly proud of their hospitals. I can't think of anywhere else I've been where the default response to an ankle twist was to go immediately to the hospital.

So we went. It was full dark by the time we arrived. The halls were sporadically under-lit by naked bulbs and fluorescent tubes, the walls and furniture crumbling. We were directed to a waiting area, and my wife was seen ten minutes later by a pair of attentive young doctors, who wrapped her ankle and advised her on care. No one asked us for money or anything else. We were back at our guest house half an hour after we'd left.

This is how Cubans are with what little they have, the things that might be of use to us. Take it, as much as you need. Don't hesitate, don't be shy. Just go.

After my wife was settled in bed, I returned to the hospital. I had a bag with me, our final gift to the Cuban people. My Spanish was so weak and disjointed, the people in the entry hall thought I was a bit unhinged at first, but once I showed them what I had in the bag they found a doctor upstairs to sit with me and register the donation. One large bottle of acetaminophen. The six remaining pills in a blister pack of a dozen decongestants. One bottle of gas relief drops for infants. Two-thirds of a twenty-pack of ballpoint pens. Everything we had left. The least we could give a hospital that offers aid to anyone who shows up on its doorstep but can't reliably purchase its own medication.

The next time we go to Cuba, we'll bring much more, and I'll remind myself to tip better. I've decided on a CUC $10 minimum. It's only fair.

CALGARY RECONSIDERED

The Walrus, June 2012

I was never going to live in Calgary, Alberta. I was never going to like living in Calgary, Alberta. I was never going to be involved in the politics of Calgary or anywhere else. I was never going to write about any of it. And yet within months of this piece being published, I was nominated as the Green Party candidate for a federal by-election in downtown Calgary—and came closer to winning than any Green candidate in prairie history.

A journalism career—not unlike any life worth living—is full of contradictions to notions that were once (often very recently) held fervently. One of my strangest and most rewarding journeys was one I took day after day down the same routes through the place I call home.

CALGARY IS YOUNG

I FIRST CAME TO KNOW CALGARY, as many do, driving along highways and broad suburban avenues, through the nowhere geography of its ample suburban periphery as my future wife toured the non-sights of her childhood. We would sometimes meander around its inner-city precincts as well, and she'd grow steadily more agitated behind the wheel as she pointed out all the spots where the old buildings used to be. It's a common tic, actually, in the coded language of old-hand Calgarians: first you trade the names of the suburban neighbourhoods where you grew up, then you offer driving directions by way of landmarks that no longer exist.

It must have been 1999, because I remember passing the open field that used to be the Calgary General Hospital, the one whose demolition Ralph Klein had triumphantly overseen the year before. We were confirmed Torontonians then. My wife was certain she was never coming back to live in this denuded landscape. We didn't see much to convince us otherwise; this was a city that blew up its hospitals.

We were wrong on many counts. We bought our first house in one of Calgary's oldest neighbourhoods in 2003, and we soon discovered another kind of city entirely. Maybe it was a matter of timing. Looking back, I think we landed in Calgary at a pivot point, a break from a century of history too rigidly defined by Cowtown iconography and the boom-bust whiplash of the oil and gas business. Calgary was always more interesting than its one-dimensional reputation, but at some point not too long ago—booming as never before, its population roaring past one million with a bullet—it became a city in full.

Calgary looks ever forward and often moves as fast as a prairie storm; its official motto, adopted in 1884, is a single propulsive word: "Onward." It can seem, at a glance, like a place with no past at all. By world standards, and even by Canadian ones, this isn't much of an overstatement. To say that it is a young city is accurate demographically—its median age, 35.8, is the lowest in Canada, and its population has grown faster than any other in the country since 2001, as legions of young job seekers poured in by the tens of thousands from Regina and Mississauga and St. John's—but it is equally true on a historical scale. In 1882, the year Sir John A. Macdonald founded the Albany Club in Toronto, Calgary was a collection of tents and shacks in the shadow of a North West Mounted Police outpost, still waiting on the arrival of the Canadian Pacific Railway. Montreal built its first skyscraper, the New York Life Building, fifteen years before Calgary got its first telephone. At the end of World War I, Winnipeg was a booming industrial city of 165,000; Calgary would not reach that benchmark until ten years after World War II ended.

The city has done almost all of its significant growing in the postwar years, made big and rich and brazen by the abundant fossilized wealth of the Canadian West. It reached full urban scale only in that sainted quarter century after the war, when energy was cheap and seemingly limitless: the apex of the oil age, lit by buzzing neon and warmed by natural gas, a plastic-fantastic space age of split-level expansiveness and shopping mall abundance and fast-food convenience, all of it delivered by miraculous hydrocarbons. And Calgary has gone from big city to aspiring metropolis in the oil age's long twilight of heightened booms and foreshortened busts. Its population today is about 1.2 million residents, roughly half of whom arrived after the closing ceremonies of the 1988 Olympics.

Calgary is, before anything else, a young city. Its peers are not Toronto or Chicago or New York, certainly not London, Paris, or Berlin, not even cities like Denver and Kansas City, which bear certain vague resemblances and share overlapping histories. No, Calgary's true cohort is that far-flung constellation of gleaming new cities that emerged from mud and swamp and flat, empty space to make the oil age manifest and global. Phoenix, Dallas, Singapore, Dubai—these are its true peers. And among them, it dwells in the usual Canadian sweet spot of peace, order and good government, between the extremes of chaos and over-planning occupied by other automotive age boom towns.

Youth is a mixed blessing, for cities as well as people. It gives a city wide-eyed optimism, boundless exuberance and a thirst for risks, but it can also lend a reckless, petulant character. The stereotypical Calgary—that land of cattle ranchers and oilmen, of easy money, big trucks and expansive sprawl—speaks to both sides of its youthful nature.

Even if you love the city deep down, you sometimes feel as if you're merely putting up with it, waiting for it to grow all the way up and become what it pretends to be. Calgary is an overnight millionaire fresh from the sale of a gas exploration company, complaining about the greed of all those farmers who jacked up the lease rates. Calgary is the home riding of the prime minister abutting the home riding of the premier, and still insisting that it doesn't get a fair shake in Ottawa or Edmonton. Calgary is the highest per capita income in Canada in a province with no sales tax, indignant that its property taxes are going up. Its conservatism sometimes scans as a youngster's I-got-mine insolence. Its emerging power and prominence come across from some angles as pure teenage bluster.

I think I first heard it said about George W. Bush that he was born on third base and reckoned he must've hit a triple. Maybe this is an oil executive's delusion, because the same misapprehension pervades Calgary's business culture. There is a tendency to see the city's (and the province's) prosperity as the manifestation of native ingenuity and gumption, rather than a natural gift, a stroke of paleogeographic luck. A vein of triumphant smugness runs through the office towers, accompanied by a mild condescension that explodes to the surface whenever its assumptions are challenged too hard. Its business leaders are masters of the over-articulated exhale, the swirling eye, and the slight, dismissive smile.

These are not particularly kind words for a city I profess to love, but I got them out of the way first because there is some truth to them, and because this is the image of itself that Calgary has projected most vividly to the rest of the country and beyond. All those yee-haws and yahoos. *Let the eastern bastards freeze in the dark* and *The West Wants In.* The social calendar dominated by a rowdy annual party with a rodeo theme, in the shadow of a hockey rink shaped like a saddle. The prime minister was in China not long ago promoting trade, but because his riding is in Cowtown he was accompanied by a guy in a plush cartoon mascot's costume with an equine countenance, clad in a shirt of tablecloth red-check plaid and sheepskin chaps and answering to the name Harry the Horse. (Did Trudeau bring Bonhomme with him to Beijing?) The first superintendent of Fort Calgary, setting a certain tone, named the place after himself without permission; it was Fort Brisebois for a spell, until cooler heads prevailed. This is the thing about headstrong kids: even the ones poised for greatness are still kids, possessed of fevered brains in manic frames, squandering so much of their overflowing energy on nothing much at all.

To be young, though, is also to be free and unfettered, eager to invent and experiment. Calgary is an energetic city, bound by few traditions and even less old money propriety. Its boosters—and it is nothing if not a city of boosters—like to talk about how their city is an entrepreneur's paradise, a place for ambitious strivers with big ideas. They probably say it too often, but it begins from a cluster of truths about the city's admirable social fluidity and the permeability of its power structures. You can arrive at the start of a decade planning to leave in a year or two, and find yourself treated like a pillar of the community by decade's end. I can say that with certainty, because I lived it.

Calgary wants you to know, maybe a little too keenly, that it's ready to take its place at the world-class table as a city of global import. Coming through the airport the other day, I passed under a banner adorned with the city's new marketing tagline: "innovative energy." It's a meaningless phrase, a multi-stakeholder committee's overwrought idea of catchy, a slogan better suited to a consulting firm. (The old one, now defunct, was "Heart of the New West"; along the way, Hidy and Howdy, the cartoon bears who had served as Olympic mascots, were unceremoniously removed from the city's welcome signs.) The rebranding campaign was years in the works. A big shot LA marketing firm had been somewhat controversially involved, and there had been hand-wringing and breast-beating in the op-ed pages about losing the priceless "cowboy brand." And, to be sure, the white Smithbilt hat and the yee-hawing cartoon bears are far more distinctive than "innovative energy." ("Where do great ideas fly?" the sign inquires. The answer—"calgary"—comes in muted lower case, which doesn't fit at all. Calgary is many things, but it's pretty much never understated.)

It's all kind of silly, this hunt for a single phrase to sum up a city of a million distinct souls, but between the lines you can hear the place trying to talk about another kind of youthful exuberance that doesn't need to holler in cartoon cowboy slang. Despite the cowboy hat bluster, Calgary doesn't know exactly what it is yet, and so it can still be shaped. And that, I can attest, makes it an exciting place to stake one's urban claim.

CALGARY IS WILD

IN THE MONTHS BEFORE we moved into our first Calgary home, my wife and I camped out in her father's basement, deep in southeastern suburbia. I would drive her downtown to work most days, and I'd sometimes take different routes back, hunting out the city's less beaten paths, trying to make them my own. One morning, as I waited at a traffic light in the heart of downtown, a flock of Canada geese went by, flying in formation not much higher than my windshield. It felt like a portent.

Calgary is a wild city. The rivers that shape its contours, the Bow and the Elbow, are not wide, silty aquatic highways, but fast-flowing mountain streams that bulge with frigid meltwater each spring and trickle down to bare rock by summer's end. There's a weir on the edge of downtown, just below the confluence of the two rivers, that disgorges disoriented fish, especially at the height of summer, and so for a couple of weeks in July flocks of white pelicans take up residence on a nearby sandbar and feast. (My father-in-law calls this scene "the sushi bar.") Farther along, the Bow and a tributary, Fish Creek, form the forked backbone of a provincial park contained entirely within the city limits. The park is an oasis for deer, wolves, beavers, and owls.

My daughter's first glimpse of untamed nature was there, on a spring hike along the banks of the Bow when she was barely three. We came upon a young bald eagle, midway through the disembowelling of a smaller bird. We watched in rapt attention, my daughter just a little anxious, but the eagle paid us no mind at all.

One fine day last fall, a cougar took up residence in a tree near the Calgary Rowing Club, on the banks of the Glenmore Reservoir, the source of the city's drinking water. It took the police, several wildlife officials, and a tranquilizer gun to dislodge the big cat and return it to the wilderness west of town.

Calgary is wild, and this is a big part of its allure: the vast network of pathways for biking and jogging along the riverbanks, the Rocky Mountains on the horizon, the western suburbs spilling into the foothills, and the eastern boundary giving way to wide prairie and fossil-strewn badlands canyons. The rural wilds feel close, just up the block or down the road. At the height of summer or on a snowy winter weekend, it can seem as if half the city has emptied out to Banff or Kananaskis or the houseboats of the Shuswap. Calgary is a city of skiers, hikers, rock climbers, mountain bikers, fly fishermen. On hot summer days, the Bow fills with an ad hoc flotilla of fishing boats, canoes, and inflatable rafts tethered together like bobbing patios, complete with beer coolers and sound systems.

If you're looking for Calgary's newest money, you'll find quite a lot of it in the hills west of the city limits, in enclaves like Bearspaw and Springbank and Elbow Valley. There are few homes smaller than 5,000 square feet out in those hills, and virtually no garages less than three-car. Broad pseudo-ranch houses and many-gabled starter castles have been strewn carelessly across the ridgelines on acre plots, their picture windows facing west toward the mountains. The brass ring is a country estate half-disguised as a suburban split-level with a boat trailer or an RV the size of a tour bus parked alongside, and a half-hour jump on mountain-bound Friday-afternoon traffic. Failing that, a case of Pilsner and a rubber dinghy on the river will suffice.

Calgary was never actually the Wild West, at least not in the gunslinging sense. For all the romantic nostalgia surrounding that first handful of pioneering ranchers, it was settled primarily by the North West Mounted Police and the Canadian Pacific Railway. Until the Mounties and the trains came, there was no town to speak of. And because the settlement was young even then—younger than Edmonton, which had been a bustling fur trading post for decades, younger than the NWMP's primary southern Alberta post at Fort Macleod—it never did become the administrative centre of anyone's map. It was neither the provincial capital nor the main outpost of the federal government back east. It had no established religious power structures, no Loyalist cliques. In Calgary, as its current mayor likes to say, "No one cares who your daddy was," and no one ever really did. If you can talk your way into the room, you'll get a hearing, and you might even walk out in charge of something new.

Back in 1912, a smooth talker named Guy Weadick, an eastern city boy who'd learned trick roping and made a living pretending to be a frontier cowboy, came riding into town and convinced four local bigwigs to throw some cash at his travelling circus, and 100 years later Calgary still organizes its summers around the Stampede he sold the city on. Weadick launched an institution and gave the city its founding myth, forever wedding young Calgary to its Cowtown reputation. It is a myth so concise and enduring that even today it continues to obscure the reality of a sophisticated, business-obsessed, hyper-modern city. Even the true nature of this Weadick fellow and his vaudevillian rodeo show have been subsumed by it.

Weadick was by all reports a sincere and charming man, a gifted trick roper and horseback rider, and a passionate, persuasive sales-man. A 1952 glossy history of the Stampede describes him as "a long, lean cowpuncher with a quick nervous walk and a Wyoming drawl." If this sounds like a character in a 1950s Western, that's because it mostly is: Guy Weadick was born in Rochester, New York, and later spent time in Wyoming and Winnipeg. By the time he arrived in Calgary, the American frontier had been closed for decades, and even nostalgic Wild West travelling shows were on the wane. The inaugural Stampede was conceived as a one-time spectacle, a grand finale.

Weadick moved the Stampede to Winnipeg the next year, but the show didn't catch on there, and so he brought it back to Calgary, where it became an annual event in 1923. Even then, it included many of its current trappings: a parade, a midway, horse racing, free cowboy breakfasts, and several novel rodeo events. But those early

Stampedes were varied affairs. There were Model T races and city tours, North West Mounted Police on horseback and Hudson's Bay Company factors on foot; photos from the first Stampede parade reveal bowler hats and Sunday finery outnumbering cowboy hats and spurs.

The "Indian Village," a sparsely attended back corner of the grounds nowadays, was perhaps the most unique Stampede feature and its most authentic frontier experience. At the time, Alberta's First Nations were forbidden to leave their reserves except on short-term permits, and they couldn't wear their traditional dress or practise their cultural and religious beliefs, on or off the reserve. The local Indian agent initially refused to allow his wards to attend the first Stampede, because it required too many days off the reserve, and he was only overruled after two of Calgary's most prominent politicians, Senator James Lougheed and future prime minister R. B. Bennett, petitioned the federal government. The First Nations pitched teepees and camped out on the Stampede grounds by the thousands, embracing the event as a rare chance to express their smothered cultures; the Stampede represented the only opportunity most Calgarians ever had to interact directly with their Indigenous neighbours.

Notwithstanding the mythmaking overstatement, there was an authentic wildness to the city that birthed the Stampede, a frontier gambler's taste for risk that lives on within its glittering office towers. It's no coincidence that it spawned WestJet and Bre-X Minerals (the positive and negative poles in Calgary's freewheeling business culture), and I still can't figure out how Norman Foster managed to convince the city's largest gas drilling operation to commission Canada's greenest skyscraper, but any day now they'll cut the ribbon on the Bow building, Encana's employees will move in, and it will replace the Calgary Tower as the city's postcard icon.

Wild Calgary changes almost as quickly as the weather, and prairie storms tend to come on with little warning. The sky was blue last you checked, and then it went purple-black and the air seemed to pulse with an electricity you could almost see, and ten minutes later hailstones the size of golf balls have laid waste to the tomato plants you tended so foolishly all spring. The only thing to count on, when it comes to Calgary, is that whatever the current weather, it will change—often quite dramatically—any minute now. My wife will tell you she's seen snow in every single month on the calendar. Some years, winter turns to summer over a single weekend, with nothing in between that would be recognized elsewhere as spring.

You never forget your first chinook. It's January or February, the darkest depths of winter's deep-freeze. You look west toward the Rockies on the horizon, and a wide arch bisects the sky. In the foreground, dark, flat cloud; in the distance to the west, bright sky and warm air pushing winter out of their path like the screen wipe in a movie. A morning that started at minus ten finishes the day at plus fifteen, the chinook turning winter to full spring in a couple of hours and prompting the Ship and Anchor pub on 17 Avenue to open its patio for the afternoon. The sudden shift in air pressure feels as if it's pinching the eyes tight even in heads not prone to headaches. It's a weekend furlough from winter, boomtown weather, and it induces a kind of delirium. Don't think about tomorrow. Make the most of it for as long as it lasts. Because the winter, everyone knows, will be back soon enough.

CALGARY IS A BOOMTOWN

IN 1912, THE CPR CHOSE Calgary as its regional maintenance depot, igniting the most intense real estate boom Canada has ever seen. Within months, there were two real estate offices for every grocery store, and more than one in every ten adult males was part of the property hustle in some way or another. The volume of building permits issued in 1912 would not be equalled again until 1949. In 1912, the population was a little more than 40,000, and so a handful of local businessmen got together and formed the 100,000 Club, with the goal of hitting that target by 1915. A few months later, as the boom built to fever pitch, they changed the name to the Quarter Million Club.

Their optimism was understandable; Calgary was exploding from the prairie earth with staggering speed. Its first permanent City Hall and its first library opened their doors during that inaugural boom, and the Hudson's Bay Company and the *Calgary Herald* completed big stone edifices downtown. When Guy Weadick's Stampede debuted that summer, a crowd nearly twice the size of the city's permanent population attended its parade. Meanwhile, in a pamphlet entitled "Calgary: The City Phenomenal," some anonymous local booster wrote, "What of the Future of Calgary? Can it be doubted? NO! Calgary's future is assured ... Nothing but a catastrophe can even temporarily impede her progress. The 'cow town' of YESTERDAY, the City of 55,000 TODAY, will be the big Metropolis of the Canadian West TOMORROW."

Like all speculative bubbles, Calgary's real estate frenzy collapsed nearly as quickly as it had come on, and the bust was in full effect by the dawn of World War I. Still, the 1912 boom set the tone for much of Calgary's long-term growth. Writer Aritha van Herk once called her hometown an unpredictable "binge city," and that about sums it up. An oil strike to the south in 1914 instigated the next boom, and the much larger discovery at Leduc in 1947 saw the Quarter Million Club's ambitious goal finally reached, as the population doubled between 1948 and 1958. OPEC price shocks set off another boom cycle, with natural gas drilling province-wide and the first bitumen mining operations up north, which busted against the dual austerities of the loathed National Energy Program and the plummet in global oil prices in the 1980s. The most recent boom seemed to ramp up in hiccupping stages of growth and decline, starting in the early 1990s and building until the crazed, full-throttle days of $150-a-barrel oil in the summer of 2008, and the vicious crash along with the rest of the global economy later that year.

Calgary lurches along on this whiplash ride whether it wants to or not. Wrecking balls swing and towers rise and excess becomes *de rigueur*—one local restaurant was rimming its martini glasses in gold dust circa 2007—and then the construction cranes abruptly fall idle and a certain number of temporary Calgarians pack U-Hauls for the long drive back to Brandon or Cape Breton. The bust of the early 1980s was particularly harsh, and it coloured my wife's impression of the city for many years afterward. The worst years were accompanied by punishing drought; the suburban neighbourhoods of her youth were full of empty windows marking lost fortunes and dead brown lawns you weren't allowed to water. Many of the families that remained clung to homes worth half their purchase price, as engineers took on pizza delivery gigs to make mortgage payments they could no longer afford.

The rise and fall of boom town fortunes take a toll. They lend Calgary a transient quality, a reluctance to dig in all the way, a sense that even office buildings and schools are temporary structures. Too many residents are unsure how it was they decided to come or how long they plan to stay. There sometimes seem to be as many Calgarians building retirement homes in Kelowna or Victoria or planning moves back to the family plot on the Nova Scotia coast as there are tending the city's civic institutions. Businesses that have been on this or that block for generations vanish overnight as retail rental rates skyrocket, to be replaced by high-rolling design stores or luxury dog accessory boutiques that are themselves gone in six months. Like the warm, welcome chinook, the good times seem to end too quickly, the aftermath of the last bacchanal barely tidied up before the next one arrives.

Amid all of this churn, it's understandable that Calgary has hung on to those institutions that seem least affected by the upheaval. Business, whether booming or busting, sets the tone for the broader civic culture. Developers come to dictate the agenda at City Hall. Corporate donors determine just how robust the arts scene will be. The safest jobs are with the CPR, which still employs more Calgarians than any given oil company. The white cowboy hats and Wild West vests worn by volunteers for the Olympics will do just fine for the greeters at the airport. And though it lasts for just ten days, the Calgary Stampede, so it's presumed, will now and forever be the city's proudest symbol: its trademark, its iconic brand, its heart and soul.

CALGARY IS (STILL) COWTOWN

THE STAMPEDE AS WE now know it—and the Cowtown monomyth it reiterates and reinforces each summer—codified itself only in the '50s and '60s, against the backdrop of a cartoonish Western revival on television and movie screens across North America. Hollywood cowboys became the preferred Stampede parade marshals in those years (Tonto, the Cisco Kid, and the Virginian all played the role), and the more varied traditions and symbols of earlier Stampedes were mostly buried under a kitschy flood of six-guns and spurs and frontier town facades. "Since the 1960s," historian Max Foran writes, "the Stampede has focused primarily on the generic western myth … The Canadian West has largely disappeared from the Stampede."

Somewhere along the way, a novelty introduced by football fans during their pilgrimage to Toronto for the 1948 Grey Cup—the distribution of bright white, wide-brimmed Smithbilt cowboy hats, made in a local warehouse run by a Belarusian immigrant named Shumiatcher, who anglicized and genericized his name for the company's trademark—became a civic tradition, a year-round reminder of the Stampede's cowboy iconography. When members of the royal family visit (Andy and Fergie in 1987, Will and Kate in 2011), they get white-hatted. When Ralph Klein went to the New York Stock Exchange to celebrate the IPO of a local company, he white-hatted the overseers of the world's most powerful stock market. When world leaders descended on Calgary for a G8 meeting a few years back, they all got white-hatted (though Tony Blair reportedly refused to wear his, and Jacques Chirac was visibly disturbed by the ritual). The day my wife and I got married, white hats went onto the heads of half a dozen guests at the reception.

Calgary can sometimes seem monolithic, one-dimensional, a lone cowboy forever in the same white hat, yip-yip-yahooing from the oil patch. Cowtown, yes, and nothing more. The sea of Conservative blue that covers most of the city on each federal and provincial election night hides the fact that in many ridings at least half the electorate voted for another party. (In recent years, the Green Party has achieved some of its best results in Calgary.) The girth, power, and prominence of the oil and gas business, direct or indirect employer of one in seven Albertans, often overshadows the fact that the majority of Calgarians work in businesses with no connection to fossil fuels. The brazen braying that erupts from some of Calgary's best-known and most heavily amplified mouths—Ralph Klein and Stephen Harper, Ezra Levant and Barry Cooper, Rod Love and Danielle Smith—tends to drown out the strident rebuttals from other local voices. (Put another way, it's better known that Calgary West MP Rob Anders accused Nelson Mandela of being a terrorist on the floor of the House of Commons than it is that his own riding association has twice tried to oust him as its nominee.)

Even Calgary's civic structure reinforces the monomyth. It is the only large municipality in the country that encompasses not just the inner city, but all of its suburbs. By reputation, Calgary is the red-neck, pickup-trucking sprawl capital of Canada, while Vancouver is the greenest city that ever bike-laned its way to the farmers' market—and you would see those stereotypes broadly confirmed by a statistical comparison of the City of Calgary (all 1.2 million residents of the metro area) and the City of Vancouver (just 600,000 enlightened souls in a metropolis of at least 2.3 million, depending on how far up the Fraser Valley you go before you stop counting). Compare metro areas, however, and you'll uncover my favourite statistical pair, the one indicating that the proportion of commuters who get to work by foot, bike, and public transit each day in the Greater Vancouver Area is just 2.2 percent greater than in Calgary.

The fine details and hard data don't amount to much, though, when the symbol is so powerful, so enduring, so widely displayed and frequently reinforced. Which is why in some circles, the mono-myth inspires a neglected younger sibling's deep, inchoate loathing. If Stampede was just Stampede, a ten-day summer festival with calf roping and fireworks and those addictive mini-doughnuts, a free pancake breakfast in the parking lot of the nearest mall, and some overzealous boozing with co-workers—if that's all it was, it wouldn't inspire such spite. But it isn't just that. It won't just stay there on the flat land below Scotsman's Hill, won't keep quiet after the last explosion in the Grandstand fireworks show. You can't just take it or leave it while the carnival is up and running. It insists on being everything to everyone everywhere, Calgary by proxy, the default iconographic setting for any discussion of the city and the province and the Canadian West in general. And to question its value, to argue that Calgary is a much more interesting city than its monomyth, is tantamount to blasphemy.

In the months before the 2011 Stampede, some mildly heated debate about party tents took place in my neighbourhood. The previous year, the white-linen Italian restaurant at the end of my block had turned its parking lot into a satellite Stampede party venue. Behind high temporary riot fencing, a white party tent went up. It was said on my block that the restaurant's proprietors had promised a staid corporate affair, but instead it had been a rollicking dance party, a noisy, throbbing, drunken mess. Which, of course, is business as usual at Stampede—except that my neighbourhood is on the far side of the central business district and across the Bow River from the Stampede grounds, and it had never before been a Stampede party venue.

The controversy over the renewal of the restaurant's temporary tent licence hit the press, and former alderman Ric McIver used his *Calgary Herald* column to explain why my neighbourhood's distaste for Stampede rowdiness was downright un-Calgarian. "Calgary's history is marked strikingly by her pioneers," he wrote. "They were the people who farmed, ranched and homesteaded during difficult times on the Prairies. They worked hard and they played hard. With no ballet to attend or white-table-cloth restaurant to while away the evening hours at, those pioneers made their own fun ... Sometimes, the pioneers would compete to see who could stay on wild horses for the longest. Sometimes, they would see who could catch the calves and tie them down for branding in the least time. Sometimes, they would see who could pack up the wagons and get back home the fastest at the end of a long work day. Over time, these competitions developed standard rules, and as a group, were referred to as rodeo. The pleasures available included enjoying a drink of whisky at the end of the day and listening to music played by whoever was available to sing and play the piano or guitar. These roots of our city are celebrated all year long, but particularly during the Calgary Stampede. The biggest and rowdiest of celebrations often take place in tents."

This is a particularly fervent defence of Calgary's monomyth (a few sentences in, you can practically hear the loping notes of "Happy Trails" on some cowpuncher's guitar), but it is far from historically accurate. Aside from a few enterprising homesteaders and land speculators, the first Calgarians were mounted police and railway workers. Samuel Shaw, one of the most prominent ranchers of Calgary's founding era, was an English gentleman of means who arrived on the prairie with a prefab woollen mill ready to be assembled, and tethered his ranch house to the newborn municipality by telegraph wire so he could play chess with colleagues far afield. The chuckwagon races McIver alludes to were a whole-cloth creation of the nimble mind of Guy Weadick (reportedly inspired by then novel car races he'd seen), and the inaugural Stampede was the first rodeo ever staged in these parts. Yes, there were cowboys on those first Alberta ranches, but their lives bore only a passing resemblance to the Old West camp-fire scene McIver describes. Even noted Alberta storyteller Grant MacEwan, who loved a good yarn more than the dull, unvarnished truth, could say only that local cowboys "rode to the sprawling set-tlement built around Fort Calgary and spent their wages on supplies and such entertainment as a community under the watchful gaze of the Mounted Police would afford."

I shouldn't be too hard on McIver. Like a great many zealous defenders of Calgary's enclosed conservative tradition (the prime min-ister is another), he only arrived as an adult, at the age of twenty-three, transferred from Woodstock, Ontario, to work in sales for Schneider Foods. You come to Calgary fresh out of school for a good job in a well-appointed office, you buy yourself a pair of cowboy boots and a wide-brimmed Smithbilt, and you accept the Stampede's marketing brochure as the true history of your adopted hometown. This, much more than any ancestral connection to the province's ranching past, is the common tradition of Calgary's contemporary citizenry. They are corporate lawyers, geologists, health care professionals, packaged goods salespeople. The money is good, the air clean, and the skies not cloudy very many days, and so they learn to say "yee-haw" and "howdy" and "y'all" (none of which have any linguistic connection whatsoever to the city's actual past) for ten days each summer, and defend their new-found fortunes year round with true-blue Tory zeal.

It's a tidy life, and it fits Calgary well. There's much that encour-ages its residents to stay safely penned inside the clean enclosure of its well-lit paths and well-worn myths.

CALGARY IS ENCLOSED

CALGARIANS LIVE IN A CITY divided into quadrants, the streets num-bered into an orderly grid even when the paths of the actual roadways don't hew to it. The southwest is known to be rich and well appointed (home to Stephen Harper and Premier Alison Redford), the northeast thought to be poor and increasingly dark skinned (Vietnamese is the default strip mall cuisine city wide, and the South Asians work all the joe jobs at the airport).

Calgarians of every quadrant reside, more often than not, in homes built with their front porches out back and their garages facing the street, and they drive downtown into multi-level parking lots each morning, and take their lunch in a vast labyrinth of a shopping mall that winds its way through the bottom two floors of every other office tower. It's an indoor routine, walled off and fenced in and carefully climate controlled. It's another fine local tradition, actually. Like Sam Shaw with his chess games, like a fellow named Riley so homesick he gave all the streets in my neighbourhood the most English names he could think up (Kensington and Gladstone and Oxford), today's Calgarians live in mild, prosperous denial of where they are and how it came to be. Small wonder, then, that the signature piece of downtown architecture is a series of enclosed walkways fifteen feet above the exposed streets below.

The Plus 15 network is a curious ecosystem, even vibrant in its way. More than sixty-two skywalks connect the downtown buildings in a network of concourses and mezzanines said to stretch for eighteen kilometres. There are back alleys of a sort, little dead-end passages lined with generic cafeterias with names like Lunch Today and Snack Rack, but the core of the network has been renovated extensively in recent years, re-tiled and skylit into a welcoming corporate retail space. The Plus 15s are an entire downtown core reimagined as a shopping mall. There are buskers, street people hunting through garbage cans for returnable containers, underused little outdoor patios, a stock ticker. It is not uncommon to see dense clusters of power-suited executives in the food courts, hammering out multimillion-dollar oil deals as they drink pop through plastic straws. It is a singular place, generic and yet wholly Calgarian, so much so that it became the setting for one of the most viciously satirical movies ever made about Calgary—without the city itself even being named.

Gary Burns's *Waydowntown* (2000) tells the story of a group of young corporate cubicle dwellers who engage in a bet to see who can endure the longest without going outside. Burns's Plus 15 network requires one small fiction—that residential towers are also part of the system—but it is otherwise a naturalistic portrayal of Calgary's business world. He shot it on location in the Plus 15s using handheld digital cameras and open sets; the passersby and background details are documentary in nature. He conceived the project as an indictment of the Plus 15 system and the city whose streets it denies. "I really thought the Plus 15 was a total disaster," he told me recently. "I was so angry about it."

He broadened his critique of Calgary in *Radiant City* (2006), co-directed by beloved local CBC radio host Jim Brown. A quasi-documentary about life in the suburbs, it switches interchangeably between settings in southeastern Calgary and elsewhere in North America. Calgary again goes unnamed; the implicit argument is either that Calgary is the ultimate expression of the suburban form, or that everywhere is at least a little like Calgary. The title comes from a 1935 book by the seminal French modernist architect Le Corbusier, whose work served as the inspiration for a significant swath of modern suburbia, in which he imagined a model city of skywalks and symmetrical glass towers. Calgary's enclosed downtown might represent Canada's most robust expression of Le Corbusier's vision—one that nearly came to full and disastrous fruition through an ambitious master planning process in the 1960s.

The proposed plans imagined Calgary as little more than a constellation of suburban enclaves linked to the downtown core by broad ribbons of asphalt and concrete. The most startling feature was the Bow Trail, a parkway six or maybe even ten lanes wide, hugging the south bank of the Bow River from one end of the inner city to the other; the CPR tracks were to be rerouted to snake alongside the new highway. To do this, Chinatown would have to be razed pretty much in its entirety. Because of its location on the south side of the river, the Bow Trail would have roared roughshod across historical residential neighbourhoods on the east and west sides of downtown, and would have involved erecting a permanent barrier of fast-moving traffic between the business district and Prince's Island, a green oasis in the heart of the city.

Even at the very peak of Calgary's automotive heyday and the height of its first great oil boom, residents vociferously rejected the plans. Opposition was particularly fierce in Chinatown and Inglewood, the east side neighbourhood that had grown up across the river from Fort Calgary, on the original townsite. Pieces of the plan, including a truncated Bow Trail on the west side, were built, but even wild, car-loving Calgary realized there were limits to how much a city should sacrifice to commuter traffic's exigencies. It was perhaps an expression of the less celebrated side of its enclosed conservatism, a level-headed prairie practicality to balance out the yahoos.

In the long term, the most significant impact of the plan might have been that it carved out a space within the enclosure to talk about traffic, which a quietly enterprising Dutch immigrant at Calgary Transit by the name of Bill Kuyt somehow leveraged into a transformative conversation about public transport. Under his guidance, the city spent the 1960s investigating everything from subways to monorails and space-age personal rapid transit pods hailed at the touch of a button, before settling on the eminently reasonable solution of light rail transit. The CTrain, Canada's first LRT system, started operations in 1981; since 2001, the whole system has been wind powered. And old Bow Trail has been a mess of construction these past few years, because the city is laying CTrain tracks right down the centre of it.

It is another of those unexpected, paradoxical things about Calgary: there is a wide range of ideas hidden inside the enclosure.

CALGARY IS THE OPEN RANGE

For almost as long as Calgary has been called Cowtown, local boosters and politicians and other assorted VIPs have argued that there is more to the city than rawhide and rodeo. Even Ralph Klein, whose white-hatted, self-satisfied mug defined its character in the national media for much of the past quarter century, can be heard making the case to the press repeatedly in the weeks before the 1988 Olympics. "There is culture beyond cowboy culture," he told reporters. "We have ballet. We have a symphony. We have theatre."

This was all true, of course, but any city with a little gumption and a lot of money can found an institution and erect a building to house it. What Calgary lacked, even in its own collective mind, was any sense of what kind of urban story those institutions were telling.

In 1912, around the time the famed quartet of local businessmen was cutting cheques to fund Weadick's Stampede idea, the big shots at City Hall were entertaining an even grander scheme. Having decided to bring in a foreign expert to help them figure out how to manage Calgary's phenomenal growth, the newborn planning department wound up commissioning a staggeringly ambitious master plan by an English urban designer named Thomas H. Mawson.

Mawson's vision has been called "Vienna on the Bow," and he cited both the Austrian capital and Haussmann's Paris as models in the plan. It proposed organizing the city into a series of radial axes around a grandiose central plaza ringed in arcaded neoclassical civic buildings: museums and vast exhibition halls, government buildings, and a palatial train station. He even mused on the possibility of an automobile elevator to facilitate traffic across a proposed two-tiered Centre Street Bridge. Had even a portion of the monumental project been built, Calgary might today possess one of the finest classical city centres in North America. But, alas, the boom went bust, as they always do, and Mawson's plan was forgotten. Calgary got its rodeo but never built a grand plaza, and utilitarian towers and prefab sub-urban enclaves became the predominant forms. Even its oldest neigh-bourhoods came to take on a suburban character.

My first Calgary home was in one of these: a 1911 stucco bunga-low in a district called Ramsay, on the working-class side of the CPR's original townsite, just across the Elbow from the Stampede grounds (close enough that the Grandstand fireworks that conclude each night of Stampede rattled our windows). Ramsay had once been an urban village abutting the stockyards, but over time it had lost much of its non-residential activity; we lived half a block from a 7-Eleven, easily the most bustling of the few remaining shops in the neighbourhood.

I became tangentially involved in local politics, and a group of us came to disagree intensely with our community association, which opposed an elaborate mixed-use development on the edge of the neighbourhood. I've stood in front of city council several times to argue in favour of a denser, more stridently urban city, so I can't remember if it was during this debate or some other one that an alderman asked me doubtfully what I thought such changes might do to the "character" of a Calgary suburb. I remember, in any case, leaving the council hearing with a dejected sense of inevitability, walking out to the parking lot on a cold, dark February evening, watching my breath emerge in thick clouds and fade with steady certainty, like a dream dissolving over and over again. The enclosure's fences would not be breached on this night.

We tired dissenters decamped to a nearby pub in a vintage fire hall, a joint called the Hose and Hound, and I spied a familiar face at one table, a local academic with an urbanist bent whom I had met at a civic engagement meeting or two. He was drinking a Coke and was hunched in cheerful, conspiratorial chat with a political consultant friend; the prof told me they were musing on who they might back for a run at the mayor's office in the following year's election. I remember thinking it was pretty bold of them—verging on delusional, really—to wonder aloud, so matter-of-factly, about something so ambitious. You couldn't even sell this town on a single piece of urban-scale development in its oldest neighbourhood without being accused of messing with its sacred Cowtown character.

A year and a half later, the political strategist Stephen Carter masterminded the campaign that got the academic Naheed Nenshi elected mayor of Calgary. (Carter went on to become Alison Redford's chief of staff, having piloted her to the premier's office as well.)

On the October night in 2010 when Nenshi was elected the city's thirty-sixth mayor, there were surprised faces all around. He was a business professor obsessed with the public sector, the son of immigrants and a native Calgarian, a policy wonk and transit geek who loved to talk about avant-garde theatre and balanced budgets in practically the same breath. He was also the underdog. He had launched his bid for the mayor's chair polling somewhere south of 2 percent support; even as that number ticked steadily upward over the weeks of the campaign, I didn't let myself believe it would top out in victory until the moment the returns said so. It was a bigger head scratcher, though, to wake up the next day and learn that the rest of the country was struck by Nenshi's pedigree. How, they wondered, had a Muslim academic of Indian ancestry captured the top political office in Cowtown? The answer, in truth, was because it never occurred to

Calgarians to wonder what any of that had to do with the guy's ability to be a good mayor.

To outside observers, Nenshi's victory seemed like the beginning of a new chapter for Calgary, but from the vantage point of his victory party at a stylish bar along what had once been brawling Electric Avenue, it felt at least as much like the end of one. There was a spirit of civic engagement and volunteerism that had been building since as far back as the famously friendly and well-managed Olympics. In the first years of the new millennium, lifestyle pieces began to appear in the national press and even farther afield, pointing out the "new cool" in a city that had found its "cultural mojo." *USA Today* ran one of these under a subhead that read, "It's hip, hopping and open for business." In the spring of 2004, the beloved Calgary Flames came within one win of a Stanley Cup, turning several blocks of 17 Avenue into an episodic Mardi Gras parade and rechristening the street "the Red Mile." The opera company staged daring original works about whisky traders and the doomed expeditions of Martin Frobisher. Its avant-garde puppet troupes toured the world. (The Old Trout Puppet Workshop, in particular, turned out brilliant Gothic comic riffs on *Beowulf*, Don Juan, and the haunted French chef Antonin Carême.) Each passing year was marked by the launch of a new music festival (local entrepreneur Zak Pashak's eclectic Sled Island festival, founded in 2007), the commissioning of a landmark piece of architecture (Norman Foster's Bow building in 2006, Santiago Calatrava's celebrated pedestrian Peace Bridge in 2008), or the founding of a new cultural institution (the National Music Centre, established in 2010). Years before Nenshi served as its lightning rod, the city crackled with wild energy.

Not long after my wife and I moved into our first home there, the city hatched a new organization called Calgary Arts Development, with an energetic, whip-smart young Saskatchewan import named Terry Rock as its founding president and CEO. The new organization's goal was, in essence, to uncover and nurture a new kind of local culture. In June 2005, Rock published an ambitious statement of purpose in the *Herald*. "From our perspective," he wrote, "the new West abounds with what historian Daniel Boorstin calls 'fertile verges' ... When the conditions are right, a verge is a crucible for creativity, a source of new ideas and connections that enable old problems to be addressed in new ways, or that simply result in new creation." He dreamed aloud of seeing Calgary declared Canada's cultural capital for 2006; he had to wait until 2012 to see it happen. During the years in between, my wife and I moved from one old neighbourhood to another, and by coincidence we now live more or less across the street from Rock and his family.

One of our favourite events on the social calendar is Folkfest—the Calgary Folk Music Festival—a four-day concert held on leafy Prince's Island a week or so after the Stampede wraps up in July. Despite the name, it is an eclectic affair on multiple stages, with some of the biggest artists from the nightly mainstage show joining their side stage brethren for intimate afternoon jams and workshops. In 2011, the top bill went to born-and-bred Albertan k.d. lang, who took the stage for the first time since she had played there in 1985 as an up-and-coming regional act. "Well, hello, Cowtown!" she announced by way of introduction. "Look at you, Calgary. You're so grown up!" The crowd greeted the sentiment with an appreciative roar. We knew what she meant. For blocks all around, the city was a study in gleaming new self-confidence.

The banks of the Bow just south of us, too long neglected, were now lined by RiverWalk, an elegantly designed, welcoming series of pedestrian paths and pocket plazas and public art overlooking the river. A bit to our west, on-site work was just getting started on Calatrava's Peace Bridge, the first truly artful span to cross the Bow inside the city limits since the Centre Street Bridge was completed in 1916. The skeleton of Foster's hyper-efficient Bow building loomed in the twilight to the southeast. To the east, a derelict district that had stymied grand schemes and best intentions for decades had been rechristened the East Village, and showed every sign of becoming a model for enlightened urban redevelopment; its anchor is the National Music Centre, which will be housed in the refurbished King Edward Hotel, a beloved old dive and blues can that had seemed destined for the city's relentless wrecking ball. Across the river to the northeast, the General Hospital site was being reimagined as a hip, European-scale mixed-use district called the Bridges. The old Grand Theatre downtown, a landmark from the very first boom, had narrowly escaped the same fate as the General, but was instead reborn as the home of an innovative company called Theatre Junction, which started reclaiming its back alley for occasional celebratory parties soon after it reopened.

Folkfest always gets remarkable weather of some sort or another: sainted sunlight and flawless evenings that fade like gentle lullabies into darkness, or else crackling lightning and a sudden, ferocious prairie hailstorm. The night in 2011 that k. d. lang sang vampy torch songs and twangy cowboy stompers was one of those perfect, sighing ones, soft sunset light and still air dancing with motes, the city glittering all around, all of us united in our certainty that there was no better place on earth to be at that moment than to be wide awake and dreaming on the open range of Calgary.

WHEN SOMEONE SAYS THEY LOVE A CITY—or hate it—they are often telling you what they think of the version of themselves they see reflected in it. They love—or hate—*who they are* in that city. I love Berlin, Seville, and New York. I came to a grudging kind of love for Delhi. I fell out of love with Toronto. I love Montreal, but don't we all, often with the adoration for a starlet on a screen, in a place you will never really belong. I've never had much time for Vancouver or Ottawa. I've hated London more than once, and I'm not sure why (but I do know it's me, not London). The hardest love is the everyday kind, the one that lingers on after commuter traffic and parent-teacher interviews and byzantine arguments at City Hall. The kind I have found, to my surprise and delight, that I now have for Calgary. It's the reason why I'm here. Still. Indefinitely.

Why stay? There are mundane material explanations, of course. Because my wife grew up in a house on an artificial lake called Bonavista in the southern suburbs, because her father is here and comes by to take the kids to the zoo, because we could buy a great little patch of 100-year-old downtown property just before the boom priced us out of the last big Canadian urban market we could afford. Because it's sunny 300 days a year, because even on the coldest day the next chinook might already be on its way. But that's not it, not all of it. That's logistics, not poetry.

I'm dug in. It still surprises me sometimes, but I am. I'm here because Calgary is a city whose best stories haven't been told too many times. Because it's a city whose best stories maybe haven't even been written yet. I'm here because everything but the cowboy hat is still an open question, wide open like the prairie, hinting on the horizon of soaring mountains. It is a young city, stupid and headstrong, brilliant and bold, and it may embarrass itself (again), but it will probably surprise you yet. It surprised me. I walk down my block in Hillhurst near the river with migrating birds overhead, I stroll under poplar branches to the century-old house with the white picket fence (no word of a lie) and the stained glass transoms over the front windows, and I cross the threshold, and I am home in a way I have never been anywhere else. How the hell did that happen?

BEARING WITNESS:
WHAT'S REALLY AT STAKE IN THE GREAT BEAR RAINFOREST

Eighteen Bridges, Winter 2011

Maybe it's an emerging truism of the Anthropocene that the most important stories will come to you randomly and without fanfare. Anyway, I was something like nine-tenths done the first draft of my book The Leap, *when an unsolicited Facebook message dragged me away from the writing for one of the most illuminating weekend trips of my life. I still don't feel like I've fully unpacked the experience—there's little in this piece, for example, about the fascinating technological spectacle of watching top-flight* National Geographic *photographers at work and recognizing, even amid the artistry, the extraordinarily narrow focus of their lens—but this was a pretty good first go at it.*

It's a tired environmentalist's cliché at this point, but there really is no substitute for full immersion in a pristine landscape to make you want to protect it at any cost—even amid the radical change of the Anthropocene.

LET'S SAY YOU'VE NEVER heard of the Great Bear Rainforest. I never had. Let's say it's a theory, a conjecture, a proper noun three words long and as real to you as fabled El Dorado or the moons of Jupiter. There it is in the subject line of a Facebook message: "invite to the great bear rainforest." A *Facebook* message, not even capitalized. Incidental. Marginal. A rumour of a place.

WHAT WOULD IT TAKE for you to care about it as much as anywhere else on earth? How long would it take to fall forever in love with the Great Bear Rainforest? It took me three days. I'll tell you how.

My afterthought Facebook message came from Tides Canada, a Vancouver environmental foundation sponsoring a documentary mission to the Great Bear Rainforest involving a handful of *National Geographic* photographers. The goal was "to raise awareness about this area, and the danger it faces from the tar sands, Enbridge pipeline and oil tanker traffic." I knew the name of the oil-and-gas pipeline developer Enbridge—I can see its downtown Calgary headquarters from my bedroom window—but I'd never before encountered the name *Great Bear Rainforest*. This would soon come to seem absurd, as if I'd reached adulthood without hearing tell of the Rocky Mountains or the Great Lakes or the Grand Banks.

I did some perfunctory research. The Great Bear Rainforest, I learned, was a protected wilderness on the remote northwest coast of British Columbia, comprising a quarter of the world's remaining intact coastal temperate rainforest. An organization called the

International League of Conservation Photographers (ILCP), a charitable group formed by several of the world's most highly regarded wildlife photographers, was midway through its annual RAVE (Rapid Assessment Visual Expedition) in the rainforest. Every year, the ILCP chooses an ecosystem in peril to document for a week. In the case of Great Bear, the imminent peril was the arrival of Big Oil. Enbridge had applied to the federal government early in 2010 to build two pipelines from an oil terminal north-east of Edmonton across 1,170 kilometres of wilderness to the industrial town of Kitimat. One pipeline would flow east with condensate (a petrochemical product that alters the viscosity of bitumen, allowing it to travel down pipelines); the other pipeline would flow west with the bitumen itself—oil, that is—at the rate of 525,000 barrels per day.

Kitimat sits at roughly the midway point of the Great Bear coast. To bring the pipeline's oil to markets around the Pacific Rim—China, in particular—mammoth supertankers would need to move in and out of a long, narrow passage known as Douglas Channel at the rate of two hundred or more per year. They would pass within a few hundred metres of the First Nations village of Hartley Bay and within a few kilometres of Princess Royal Island, home to the world's largest known population of kermode bears. The kermode is a rare subspecies of black bear with a recessive gene that renders its coat a ghostly white; locals call them "spirit bears," and they are the billboard icons of the Great Bear wilderness. The tankers would also pass by dozens of unnamed, barely explored salmon streams, the vital biodiversifying arteries of an entire ecosystem.

The *National Geographic* photographers were in the rainforest to sound an international activist alarm against the Enbridge pipeline. If I could get myself to Vancouver International Airport by the morning of 11 September, I could catch a float plane to Hartley Bay and see their work firsthand.

I thought at first it might not be worth the trip. I'd long regarded climate change as the great black trump card in the conservation deck, the overarching crisis bearing down on us with such ferocious transformative power it will erase any act of regional conservation, however noble; keeping one pipeline's bitumen from reaching Kitimat, after all, would do nothing to keep out the carbon dioxide emissions released by an oil-hungry world. If an awareness campaign isn't aimed at ending the age of fossil fuels *in toto*, I tend to see it as an act of deck-chair feng shui on the biospheric Titanic.

Still: this was uncharted territory. *Here there be serpents*, at least on the mental map of my own experience. Float planes, coastal First Nations villages, temperate rainforest, spirit bears—this was an irresistible enticement. Thankfully so: there was much to learn about the spirit bear's iconic place in the global struggle to contain the climate crisis.

ATTACHED TO THE FLOAT PLANE terminal at the airport is a welcoming bar and grill called the Flying Beaver. It's got a patio out back, an unassuming oasis jutting out over the placid little bay of Pacific water that serves as the runway for the float planes. It's a tucked-away corner of a tucked-away corner of a tucked-away corner of the main airport, and I was seated there with the dregs of a coffee when I heard the buzz of propeller engines for the third time that morning and watched an odd, boxy little airplane that looked like something Howard Hughes might've owned come down and down into the bay.

There were no floats protruding from its hull. It bellyflopped onto the water and seemed to disappear under the wake it raised, and I had an involuntary adrenal jolt as I waited for it to resurface. It came up the ramp next to the patio and crossed the access road onto the tarmac in front of the float-plane hangar, and a few minutes later its pilot came into the bar and grill and introduced himself to me as Julian MacQueen.

I'd traded emails with MacQueen, but I knew nothing else about him except that he was flying the 630 kilometres up the coast to Hartley Bay that morning. I'd imagined a plaid-jacketed adventurer type, maybe a bush pilot who ran supplies into logging camps. Instead, I climbed into the cockpit of a beautifully restored Grumman G-44 Widgeon and sat next to an affable business executive with the soft, elegant accent of Alabama gentry. MacQueen is a born-and-bred Southerner married to a Canadian, and they keep a summer home on Saltspring Island. The rest of the year, he is a hotelier, the largest single owner of resort properties on Pensacola Beach in Florida. He'd let his neighbours on Saltspring know he was open to errands of mercy, which is how he'd come to volunteer to ferry journalists and photographers in and out of the Great Bear Rainforest.

Most modern float planes are aircraft on pontoon stilts, but MacQueen's Grumman G-44 Widgeon is more like a boat with wings. As we taxied out into the bay, the water lapped against the windows at chin level. When MacQueen leaned down on the throttle, the plane's nose perked up like a motorboat's. The whole vessel rose up to skim the surface, he pulled back on the wheel, and we were airborne. Through mixed cloud and drizzle, we banked north and chugged past the Georgia Strait's inhabited islands, a patchwork of cabins and grand rural estates that grew more sporadic as it gave way to wilderness. We exchanged small talk that quickly turned meaningful, and I was reminded that dramatic characterization and deep irony were not inventions of literature but reflections of reality, such as the one inhabited just now behind the controls of a Grumman Widgeon by Julian MacQueen of Pensacola, Florida.

MacQueen loved to talk about the Widgeon, which was a smaller version of the more famous Grumman Goose. Only a few hundred Widgeons were ever made, all of them in the nineteen-forties and early nineteen-fifties, and MacQueen had hunted for one for years until he found one left chained to a tree in rural Nigeria.

Why, I wondered, would someone abandon a vintage float plane in the backwoods of Nigeria?

Oh, you know, the oil business—this was his response. Some drilling project must've finished, and the flight home was more trouble than it was worth. The world was just full of the oil industry's junk, wasn't it?

Somehow MacQueen got the thing to England, where he had it meticulously restored. It was an exquisite vessel now, a handcrafted airborne sailboat, all teak wood and analog dials and brass-coloured fittings. He called the plane his "time machine"—not because it was a throwback but because of the way it could skip from lake to harbour to welcoming bay across the continent, compressing space to the point where no place seemed far away. The Widgeon was a device to conquer time.

Below us, the landscape had grown wilder in bands, like a graded map of civilization. The sprawling holiday homes of the southern

Gulf Islands had given way to knots of more modest settlements, logging towns and native villages, and then the blanket of forest and dark water was broken only by a derelict cannery or the industrial gridwork of a salmon farm. The landscape below looked wild, but still it was not truly remote. The coarse hand of industry had scrawled its history across the steep forested peaks and blackwater inlets, telling a gruff tale of extraction, exploitation, exhaustion: Clearcuts of varied vintage like patches of diseased flesh, the age of the infection told in the height of the low monocrop trees slowly filling in the poxed land. Fishing boats splitting the water in slow surgical swoops of wake. Barges moving slow under the weight of logs. A shuttered pulp town. Fish, trees, stumps, barges, nets.

It was a map of Canadian history in miniature, a tale told most emphatically in the seizure of its natural resources, the story of a country founded by fishermen and fur trappers.

Beyond the northern tip of Vancouver Island, the clouds closed in and the land became fully wild. Steep black peaks loomed in the mist and spilled down into the darker black of the ocean. A vertiginous sense of departure swallowed the plane, a foreboding born of drastically altered scale—this tiny plane among giants—and MacQueen and I returned to our small talk. I soon realized he didn't know anything about the pipeline or the spirit bear or any other aspect of our trip. As I filled him in, his face came alive with the shock of recognition. He dug through the bag between our seats and produced an iPad.

"I've got to show you," he said, "what I've been doing the last few months."

At the start of the summer of 2010 (this was the story he told as he scrolled through photo albums on his iPad) oil from BP's massive blowout in the Gulf of Mexico began to wash ashore along the Florida panhandle. One of the first places it arrived was the beach in front of MacQueen's row of resort hotels on Pensacola Beach. Summer is the peak tourist season in Pensacola, the ninety days that sustain the business for the other 275, and the summer of 2010 slid away on an oil slick.

MacQueen showed me pictures of Gulf waves gone deep purple with spilled crude and golden sand covered in inky brown tar. Legally forbidden from disposing of it himself, he'd spent weeks documenting his losses for BP's lawyers, and then he'd cashed his first compensation cheque and headed to Saltspring to unwind from the ordeal. He had 700 employees, a vibrant business twenty-five years in the making, a million dollars in insurance paid out every year to guard against potential disasters. "The last thing on anyone's list was an oil spill," he told me.

The skies began to clear. A few kilometres south of Hartley Bay, MacQueen spied a couple of black shapes in the water below and banked the Widgeon around and down for a closer look. They were humpback whales, the Great Bear Rainforest's sentinels, signalling our arrival in a world far away from Pensacola Beach and Vancouver airport, a place that could legitimately claim to be outside that world, perhaps, were it not for the exigencies of overseas oil shipping.

My notes from this point on grew steadily more sporadic, staccato, episodic. I carried a notebook, but chose to keep my digital recorder in my bag. I had to decide whether it was more important to report on Great Bear or absorb it.

Hartley Bay is a knot of modest homes gathered on the shore of a small bay at the end of a rugged Great Bear promontory, accessible only by boat or float plane. There are no cars on its narrow lanes, and many of the homes are connected to each other only by elevated wooden boardwalks. There's a community centre, a church, and a small marina. It's home to a population of 160, all of them members of the Gitga'at First Nation. They are the only permanent inhabitants at the mouth of Douglas Channel, the proposed thoroughfare for the oil supertankers which are to be fed at the rate of nearly one a day by Enbridge's planned pipeline.

The Widgeon circled into the bay like a bird of prey, bellyflopped, and then chugged toward the pier. A young man trotted out to meet us. Norm Hann is a wilderness guide and part-time teacher at the Hartley Bay school who was assisting with the photo expedition. He was a story himself; in the spring, he'd paddleboarded the 400 kilometres from Kitimat to Bella Bella to bring attention to the threat posed by oil-tanker traffic.

Once MacQueen had docked his plane, Hann led us through the village to a handsome modern house set into the last broad ledge before the face of the hill grew too steep for habitation. He explained that it was the home of Cam and Eva Hill, a Gitga'at couple who'd adopted him as his Hartley Bay family. Inside, two young children were watching *Hannah Montana* on a flat-screen TV, and we settled into the spacious kitchen for rolls and coffee. MacQueen dug out his air navigation map, and Hann traced the proposed supertanker route. He pointed out the dramatic S-curve around nearby Gil Island, just across the water from Hartley Bay, whose rocky shoals had sunk the British Columbia government ferry *Queen of the North* in 2006. Hann ran these channels and bays often on his paddle board, and he knew the local microclimates at the scale of a single paddlestroke. Douglas Channel—the narrow inlet leading from the proposed oil terminal at Kitimat—is a natural wind tunnel, with vicious crosswinds up and down its length. Hecate Strait, the wide passage separating the Great Bear coast from Haida Gwai'i—which oil tankers would traverse once they'd navigated through the tightly clustered islands of the Inland Passage—is even worse, a maelstrom of howling hurricane-force winds and shallow, choppy seas. It is, says Hann, "one of the most treacherous, narrow bodies of water anywhere in the world."

As we were leaving, I scanned framed photos of the happy Hill family on the walls. There was Cam Hill struggling proudly under the weight of prized salmon and halibut catches, pictures of orca pods and a Hawaiian vacation. An aquarium stood by the front door, home to a single brilliantly coloured prawn. The ocean, its bounty and mystery, was everywhere. Outside, clouds hung low over the village, the incidental lives of people reduced to a delicate, transient band between sea spray and falling mist so thin it was barely there at all.

KING PACIFIC LODGE IS A luxury resort situated in a cozy bay on Princess Royal Island, just a few kilometres south of Hartley Bay. The main building is a handsome three-storey structure made from local pine, cedar, Sitka spruce and stone, with broad balconies and the faintly Victorian air of a paddlewheel steamer. Rack rates for a three-night stay in one of its seventeen rooms run from $4,900 for a standard to $12,685 for the Princess Royal Suite; it has been named the best resort in Canada in the *Condé Nast Traveler* readers' poll the last four years running.

King Pacific has its own bakery in the basement and a pantry over-flowing with house-made condiments and preserves. The water supply is drawn from a local creek, the laundry is phosphate-free, and the greywater in the septic system is scrubbed naturally by bacteria before being expelled back into the ocean. The menu skews hard toward local and fresh, and the lodge employs a number of Gitga'at through a training program in Hartley Bay.

The entire resort is built on top of a salvaged U.S. Army barge, a great slab of floating steel that is towed to Prince Rupert every fall to pass the winter in drydock. Aside from the slumbering diesel generator, King Pacific Lodge leaves no trace of itself behind.

The island behind the lodge is the quintessential Great Bear landscape. It is laced with trickling creeks and rushing salmon streams, providing home and sustenance to the largest permanent population of spirit bears anywhere. In the summer of 2009, a pod of humpbacks settled into the harbour off the front deck for three weeks, often breaching to greet newly arrived guests with a friendly spout of blowhole spray.

The daily life of the lodge revolves around the vaulted central great room. When I arrived with MacQueen and Hann, the great room's couches and easy chairs were populated with *National Geographic* photographers and a documentary film crew waiting on a helicopter ride.

Hann led us into a small reading room off the great room for an introductory slideshow. There were a few details about the lodge itself, but mainly he talked about Great Bear: About the Cetacealab research facility on nearby Gil Island, which has been tracking the extraordinary return of humpback whales to the Great Bear's waters in recent years—almost 200 of them now, five times the number seen a generation ago, when the memory of whaling boats still haunted these waters. About the more tentative but no less exciting return of the fin whales. About Great Bear's unique coastal wolves, a subspecies of timberwolf genetically distinct from their inland cousins. About the black bears and grizzlies and beloved spirit bears of Princess Royal and beyond, their numbers swelling as the salmon streams filled each year with thicker runs. About the salmon as keystone species, the extraordinary ecological bounty of the salmon runs—"the miracle of life, really," is how Hann put it—and how they provide food for more than 200 species in the rainforest, from bears and wolves to trees and ferns. About the Gitga'at, their traditional fishing camps and smokehouses, the way they tell time by the fishing seasons.

About a world in exquisite balance.

JULIAN MACQUEEN AND I were bunking on a trimaran docked at Hartley Bay, and by the time we made it back it was evening and we were on the slightly dizzy side of hungry. The boat's owner, Ian McAllister, founder of a conservation group called Pacific Wild, was hunkered over the small galley stove, pan-frying halibut steaks as thick as a Melville novel. The fish had been swimming in a Great Bear channel that morning; you could tell because the meat of the steaks curled just a bit in the pan. The aroma was intoxicating. There was wine already open. We settled around the trimaran's cozy kitchen and feasted. I knew three bites in that I'd be telling people about the incomparable taste of fresh-caught halibut in Hartley Bay for years to come.

McAllister had been working and exploring in Great Bear for years. He has published two books of photographs of the rainforest and its inhabitants, and his Pacific Wild organization, co-founded with his wife, Karen, had helped organize the ILCP photography blitz. He was serving as its all-in-one project manager, logistics coordinator and communications chief.

After dinner, he took us to a fishing boat that had just returned to Hartley Bay after ferrying photographers through the wildest stretches of Great Bear. A couple of *National Geographic* photographers were seated around the boat's small kitchen table alongside the vessel's husband-and-wife owners, looking through their best shots on a laptop. They showed us pictures of sea otters, birds of prey, humpback and fin whales, a wolf trotting along a rocky coastline with a fat salmon in its jaws. One picture captured a pink salmon as it flung itself bodily from a steep coastal stream, seemingly suspended in mid-air perpendicular to the water in a feat of physical strength so implausible it looked photoshopped.

When we returned to the trimaran, McAllister and I stood on the dock chatting. He reckoned Hartley Bay was a last stand. All or nothing. There would be a pipeline or there would be spirit bears. This, I sensed, was the project his whole working life had been building toward, maybe even a chance to shape the course of history. Jane Goodall had her chimpanzees and Cousteau his aquatic adventures, and Ian McAllister would define his career by the security of the rainforest's spirit bears.

I told him I was beginning to think that this went far beyond one pipeline and one wilderness, however vast and vital. There were two grand narratives of what Canada was and what it wanted to become—two ways forward for the whole overheating, over-exploited world, really—and they were grinding hard against each other in Hartley Bay. Everything Canada had long been was aligned on one side of the battlefield: the forces of resource extraction and global trade, oil money and the insatiable self-interest of corporate profit, exploitation, and colonial domination. On the other side was a more recent vision of Canada, one that saw itself as a responsible steward of the country's extraordinary ecological wealth, that respected Indigenous rights, that sought reconciliation and balance. The Canada of the Hudson's Bay Company, the Grand Banks collapse, the Indian Act, MacMillan Bloedel and now Enbridge stood against the Canada defined by the national parks, Greenpeace, Nunavut, the Montreal Protocol and the Canadian Boreal Forest Agreement.

For almost half a century, environmentalists, natives and loggers had fought pitched battles over their perceived self-interests among the towering trees of British Columbia, only to realize almost too late that the long-term health of the forests were in everyone's interest. As we sat there in the cool, clear night, just the two of us leaning on a trimaran, I wondered if the little village of Hartley Bay and some achingly beautiful photos taken by *National Geographic* photographers could teach the global oil industry the same lesson.

EARLY MORNING, THE CLOUDS low and foggy amid the rainforest trees, and Ian McAllister was at the helm of his trimaran, chugging steadily across the black water of the bay, bound for a salmon stream on Gribbell Island. A few days earlier he'd set up a "camera trap" there—a camera paired with a motion sensor used to gather images of the Great Bear's more reluctant celebrities—and he wanted to see if he'd caught anything with it yet. He'd brought Julian MacQueen and I along to get our first taste of the real Great Bear wilderness.

I was out on deck, watching the cloud cover burn slow and steady off the water. I turned back to watch Hartley Bay recede in the distance. The dark hills had emerged from cloud, hunched around the black bay in a semicircle like great ageless sages. The village below was still mostly shrouded, a few square whitewashed facades peaking out, an arrangement of children's playthings at the feet of the ancient mountains. McAllister joined me on the deck and followed my gaze. "You've got every major oil company in the world and the world's second-largest oil reserve looking to diversify its markets," he said, "and the only thing in the way is that little community."

We crossed a wide channel of open water and came an hour or so later to a small cove on the far shore. The shallow water was lush with fat purple starfish and clams. McAllister stepped up onto the prow and let loose a long keening howl.

He'd been hoping to get some photos of coastal wolves feasting on salmon, and he was checking to see if any were still around. The only reply came from the stream mouth as coho by the score hurried away from the sound, rippling the water. He dropped anchor and we took a zodiac to shore. He checked his camera—no significant visitors yet— and then led us on a hike upstream.

The banks of the stream were spongy like peat and so thick with foliage they seemed to exhale when you stepped on them. Grasses and underbrush were shoulder-high, taller, a Jurassic landscape of ancient, mammoth plants. And then we were in shadow, lost among the true giants. The forest was towering, majestic, impossibly alive. Moss and lichen hung from every branch and crawled across every stump and rock. The tops of the cedar and Sitka spruce around us were mere hypotheses somewhere far over our heads. The creek's trickle became a steady growl as we moved farther inland. The air grew so thick and fragrant it was less like hiking than pressing through a membrane.

The stream's banks were like an emperor's dining hall in the aftermath of some frenzied bacchanal. Every few metres, we came upon salmon carcasses. Some were bloody smears of gut and skin—the remains of a bear's feast—but most were decapitated with almost surgical precision and otherwise fully intact. The wolves came first to the salmon runs, while the bears were still gorging on berries farther inland. For reasons not precisely known—possibly to avoid some parasite or other pathogen in the salmon's bellies—the coastal wolves mostly eat only the heads of the salmon.

McAllister waded into the stream like he was stepping onto his back deck and then led us tromping and sopping up the far bank. He stopped at a heaping mound of dung just a couple of hues shy of grape Kool-Aid. "One hundred percent berry-fed bear scat," he explained. Farther along, he pointed out a smaller pile of scat, bending down close to study its contents. "Wolves spent some time hanging out here. There's river otter in their scat."

He led us back toward the stream's mouth, pausing as we emerged from the trees to admire the proportions: the band of golden grass giving way to rocky shore, the tidy semicircle of coast, steep black rocky banks and dark water and the hills of other islands on the horizon. "It's a beautiful scene as it is," he said, "but imagine a pack of wolves down there, feasting on salmon." He paused. "It could be pretty powerful."

He was sizing it up not as a Group of Seven landscape but as a propaganda poster. *Look at this—aren't you awed?* He wanted me to wonder whether there could be anything more incongruous than a supertanker the length of thirty school buses chugging past. Could you imagine, he was asking, what this might look like gone thick with black sludge? Black like the surf in front of Julian MacQueen's Pensacola hotel?

WE WENT TO ANOTHER stream up the Gribbell Island coast, one McAllister knew was a favourite bear haunt. A boat carrying a few of the ILCP photographers had met up with us in the cove, and a CBC radio reporter had left them to accompany us to this second creek. My "waterproof" hiking boots started to fail about ten strides into the hike up the stream bed, and now I was hunched behind a log at the water's edge with McAllister, MacQueen and the CBC reporter, waiting on the predators.

The river around us was mildly pitched, growing steeper by the yard and so thick with salmon in their final spawning death dance that sometimes they bumped our boots. I would select one swimming in the nearby water and focus on it, watching its body wriggle furiously as it waited beneath the next rocky rise in the stream for some unseen cue. And then suddenly the water would erupt, there would be a sense more than a sight of movement, and the fish I'd been watching was gone farther upstream. A trace memory of motion and then nothing. It was like observing a Rapture. And it was here, in the roaring silence of the churning stream, that I experienced the rainforest's singular *satori*.

I was outside of time, vanished from the landscape.

The world belonged to the rainforest, to the trees and bears and flailing salmon. I was there only insofar as I was observing it.

Our recorded Western history seems incidental and transient here. Great Bear is ancient beyond our reckoning. The frontier's final end. I had no business being here, brought by float plane and trimaran, carried on a flood of gasoline, hurtled by oil to this rush of water and fish canopied by dinosaur trees. There was a perfect balance, so rich with emergent life it seemed volcanic, and I could not hope but lessen it.

This is the transcendence of Great Bear, its first and most pointed lesson: We are too small to play among these giants. We know not what we do here. It can only end badly.

THE BEAR EMERGED FROM the forest to our right maybe ten metres in front of us, a black bear of standard hue, moseying like something in a cartoon. Everyone's sudden snap to full alert shook me from my reverie. If the bear saw us, he paid us no mind whatsoever. He stopped midstream and rooted for a moment or two among the salmon, but came up empty. He tried once more, grew bored with it, and sauntered on.

When the bear was gone, McAllister told us they rarely bothered with the salmon while berries were still plentiful. In a few weeks, though, they'd be down here gorging, drunk with fish, so stupified by their feast, he said, that you could step right up to them and practically knock them over and they'd do nothing but fix you with a sleepy stare in return. I couldn't tell how much he was exaggerating.

Our attention turned again to the water. There were mostly pink salmon in this stream. They were all around us, fighting with all their strength to move another body length upstream, their scales falling away to reveal raw flesh as the struggle exhausted them to death from the outside in. Yet it was strangely tranquil for all the commotion, a ritual whose intent was so singular and self-contained it felt far away even as you stood in the middle of it.

McAllister explained how little of this we really understand. For a very long time we'd presumed this was a crude endurance match, a survival-of-the-fittest contest, with the salmon that dodged the snouts and claws of bear and wolf and made it the farthest upstream rewarded with safer spots to lay their eggs. Now that we could tag and track individual salmon, though, we'd come to realize that the fish are actually following some guidance system far beyond our reckoning, fighting to lay their eggs and then die not just close by but within mere metres of the spot where they themselves had hatched.

McAllister bent down and poked the streambed with an extended forefinger, and when he pulled it from the water it was topped with a tiny pink-white sphere the size of a peppercorn. A salmon egg. He was standing in a small eddy of clear water, and you could see the eggs scattered like confetti among the cedar-bark debris at his feet, each of them placed there by a dying fish that had only ever known the place as an infant, and which had then travelled thousands of kilometres across the years and then returned to this exact spot—improbably, preposterously, *miraculously*—to complete life's cycle.

These are just words. We talk about the cycle of life as if we invented it by naming it, not as if it were as unknowable as the cosmos or the Judaic god's true name. Formerly, I knew salmon primarily as a piece of common meat on a plate, a lifeless pink quadrilateral almost impossible to reconcile with the power of the fish racing upstream in cold autumn water in the Great Bear Rainforest, where *wild* is not a sales pitch but a way of being.

I don't think I've ever felt as irrelevant as I did standing in the middle of that salmon run.

LATE AFTERNOON ON THE deck of Ian McAllister's trimaran under clear skies, the light magical, the landscape ancient and immutable. The mouth of Douglas Channel in particular is a postcard of wild Canadian majesty, a flawless arrangement of flat black sea and low forested hills set against glacier-capped peaks on the horizon.

Now, place a modern supertanker in the foreground, and wonder at what it adds to the scene.

A typical supertanker—specifically a very large crude carrier (VLCC) or ultra large crude carrier (ULCC) of the sort that would depart the proposed oil terminal in Kitimat almost every day—is at least 300 metres long, maxing out north of 400. A thousand feet long, half as long as Hartley Bay's coastal hills are tall. At its broadest point, its beam measures more than fifty metres, wider than half a football field's length. The largest ULCCs can carry more than two million barrels of oil. A floating colossus, a self-propelled city block, a mobile reservoir: the scale is at the outside edge of most people's imaginations.

I tried to picture it, huge and vulnerable, as I sat on the deck of the trimaran, eavesdropping on an interview the CBC reporter was conducting with Julian MacQueen. They were talking about the BP spill. "I spent 25 years building up a business," MacQueen was saying, "and overnight it was gone. It was one of those gutwrenching, heartstopping events that you just don't see coming." There was a pause, another question I didn't quite catch, and then he said, "I can't tell you the feeling of helplessness that I felt."

Far on the horizon, with a timing that verged on punctuation, a plume of mist rose up out of the dark water, and then another. The water was backlit by the setting sun, the light dancing against the exhalations of a pod of humpback whales. They swam alongside us for a spell, breaching and rolling, their great pale bellies gleaming. I had to wonder how they'd greet a vessel twenty-five times the size of our trimaran, whether they'd still feel welcome in a place that was home to such leviathans.

IN THE EVENING, NORM HANN took Julian MacQueen and I to meet Helen Clifton, a revered village Elder. She lived in a tidy home up the hill from the harbour. In her foyer, a glass case held old-fashioned China dolls of native girls. She was dressed in a smart red sweater, moccasins and pearls. Ceremonial masks hung from the sloped ceiling of her living room above a flatscreen TV.

Hann called her Granny (as does everybody in Hartley Bay), and as we sat down to talk a steady stream of grandchildren and great-grandchildren came through to say good night. Clifton is a gifted storyteller, and after the kids were gone, she told us stories. She talked about the night the *Queen of the North* went down across the bay, about the months afterward waiting in vain for its toxic wreckage to be removed. She recounted a long history of mistreatment and mistrust from the Indian Act to a recent visit from a pair of Enbridge representatives, with their "mumbo-jumbo talk" of *consultation* and *reconciliation*. She talked about her own childhood in Prince Rupert, the daughter of an English immigrant tailor and a coastal native woman, the kids at school teasing her, calling her "half-breed."

Clifton worked in a Prince Rupert cannery as a young woman, and then she'd met a Gitga'at man and married him, and the rest of her life had been lived in Hartley Bay. She was interested to hear about new developments in clean energy, to talk about eco-tourism and other new avenues of training for the young people in the village. She wasn't against change; she was opposed to exploitation. She explained her stance with a sort of parable, a story about her late husband out on a hunt.

"My husband was a hunter, you know, he was trained from a little guy that there's no way that an animal would suffer, just had to shoot them in the head. And so I guess his father had told him about the white bear on Princess Royal. And because he hadn't seen it, he thought his father's just telling him this story about this bear that looks like a ghost that's in the woods—just to make him be more careful as a hunter.

"When he was about in his early fifties, he went out deer hunting, and there was this bear. And it stood up on its hind legs, and my husband said it seemed to just grow and grow until it was seven or eight feet tall. And being a hunter, he just, he brought his gun up and shot at it. And then it wiped the blood off its shoulder and it just hollered. And he just, you know, he just choked right up, because this was real. 'What my father told me—it is real. And I didn't believe my father.'

"And this bear hollered out crying, like a human. He said it was just like a woman had screeched out in hurt. And brushed at its shoulder. 'What have I done? I wasn't hunting that animal. I wasn't here for that. We have to go follow it, because if I've wounded it, then I have to kill it.' They followed it, and there wasn't the drops of blood that a badly wounded bear would have.

"And so he comes back home, and he says, 'Guess what I saw.' And he said, 'You'll find it hard to believe. Because,' he said, 'I never believed it all my life.' And he told me about this experience with a bear. 'Oh, why didn't you kill it,' I said, 'We would've had this white bear fur.' And he said, 'You foolish woman, that's your white blood talking.' So we both had to laugh. He didn't say that in an insulting way, just because, he said, 'It was only meant for me to see that it was real. Now I have to do something about it to protect it.' And so he told his people they are not to shoot that animal no matter what. 'It's there. I saw it. And nobody is to hunt it. We don't need it for nothing.'"

The spirit bear, Clifton reckoned, had visited her husband to teach him—to teach all the Gitga'at—that the bears were not there for the taking, that the spirit bear itself was the master of this place. Her husband did not stop hunting—he never doubted that the Gitga'at and everyone else needed to draw on nature's bounty to live—but he understood there were limits. You couldn't take *everything*.

When she finished the story, MacQueen introduced himself as the pilot of the float plane down at the marina. He told her about his hotels in Pensacola, about the BP spill and the decimated economy left in its wake. He wanted her to know that the Gitga'at would not stand alone in their opposition to the pipeline. "It's not just your battle to fight," he told her. "Here I am fighting it in Pensacola, Florida, and I'm up here. I mean, there's a momentum that's building up against this sort of thing. And I think there is hope, that these kinds of things can be stopped."

"I try to be optimistic," Clifton replied, "and that has to be every day. Every day I have to say to myself, you know, that there's hope."

JULIAN MACQUEEN'S TIME machine was crowded for the flight back to Vancouver. There was a press conference about the ILCP event scheduled for that afternoon, and Ian McAllister had brought along Norm Hann and a young Gitga'at wilderness tour operator named Marven Robinson. The three of them were guiding MacQueen across Princess Royal, buzzing low past dozens of streams, marvelling at a place they knew intimately, transformed by the Widgeon into something wondrous and new.

We were all wearing headsets so we could talk over the Widgeon's growling engines, and the chatter was steady and gripping, a narration of the scene below as skilled as any nature documentary's soundtrack.

"This lake is larger than I thought," Robinson said, looking down at the flawless mirrored surface of a lake tucked amid the summit peaks of Princess Royal. The lake reflected back the Great Bear tri-colour of deep green forest, golden grass and grey stone, a flag of liberation for the sovereign wilderness below. McAllister pointed out a small waterfall, and Robinson said he'd once seen five spirit bears feeding there. They traded tales of Great Bear wildlife like they were talking about their neighbours.

Just south of Bella Bella, McAllister directed MacQueen up a long, narrow tidal estuary. It was protected wilderness, the site of Pacific Wild's first successful campaign. They'd been selling hot dogs to raise the money to buy the land ahead of the logging companies when one of Warren Buffett's sons kicked in $1.5 million and secured the deal. There were a couple of humpbacks swimming up the estuary as we headed back down toward open water.

As we banked again south, Hann, McAllister and Robinson took turns pointing out the sights. A pod of humpbacks. A pale red cloud of krill in the waves. A clear cut. A salmon farm sprawled across the mouth of several wild salmon streams. A huge grey whale breaching over and over off the pristine powder-sand beach of an island off the northern tip of Vancouver Island.

MacQueen turned the Widgeon back around for another look at the grey whale, and then another and another.

"You can see wolf tracks down there," McAllister said, pointing to the beach.

"Oh, man, this is amazing," Robinson said.

"This is really, amazing," Hann said.

We all said it or something like it. *Wow. Whoa. Oh man.* What else could you say when the land and sea made the limits of language—of the human imagination itself—so clear?

I'd scheduled my flight home to Calgary too early to make the press conference, so I said my goodbyes on the tarmac at the float plane terminal. In parting, Marven Robinson told me about a conversation he'd overheard between his mother-in-law and Helen Clifton, both of them in the twilight of their lives, both vowing to make the campaign to block Enbridge's pipeline their "last fight." They told one another that they needed to teach us all one final lesson, the one the spirit bear had taught the Gitga'at years ago: *You can't take everything.*

Canada is an improbable country in many ways, and sometimes this makes our nation all the more powerful and wondrous. But it can also seem ridiculous, absurd. No: outrageous. It's outrageous that a legislature on the other end of a vast continent from the Great Bear Rainforest—in another world, really, another age—assumes the authority to decide its fate. What do they know in Ottawa of salmon or spirit bears? I spent three days in the Great Bear Rainforest, and it was long enough to know that we can presume to have no such authority, that it is not for us to decide, ever, what Great Bear is *for*.

"We have to win this one." This is what Marven Robinson's mother-in-law had told Clifton.

It's what Robinson told me in parting.

It's what I'm telling you now.

ACKNOWLEDGEMENTS

THE WORK IN THIS BOOK SPREADS across nearly the entirety of my writing life to date and involved the expert work of more editors, copy-editors, photographers, designers and publishers than I could hope to name. I'll do my best to thank everyone I can recall working with closely on these pieces and hope the ones I fail to include will forgive my memory lapses.

First, to Andrew Heintzman and Evan Solomon, *Shift* magazine's gutsy co-founders, my deepest thanks for building the platform from which I launched my career. And equal thanks as well to Felix Vikhman, who brought me into the magazine's fold. I owe *Shift* editors Laas Turnbull and Neil Morton eternal gratitude for taking repeated risks on a young writer's incoherent pitches. Joanna Pachner, Maryam Sanati, Greig Dymond, Douglas Bell, Andre Mayer, Clive Thompson and Marijke De Looze all lent editorial guidance, logistical support and/or professional wisdom to the *Shift* pieces in this collection. My thanks as well to Ian Connacher, Sheila Heti, Barnaby Marshall, Dave Sylvester, Christian Bailey, Carmen Djunko, Matthew McKinnon, Lianne George, Rodney Palmer, Steve Park, Kevin Siu, Rolf Dinsdale, Malcolm Brown and everyone else in the *Shift* clan who helped these stories or this writer in one way or another (which is basically everyone who ever worked there).

A hearty thanks as well to my second home sweet feature writing home, *The Walrus*, and especially to John Macfarlane, Jeremy Keehn, Amy Macfarlane, Sasha Chapman and Shelley Ambrose, for their editorial and publishing guidance.

Thanks to Julie Crysler, my editor at *This* magazine, and to Curtis Gillespie and Lynn Coady, my editors at *Eighteen Bridges*.

My deepest thanks to Jeet Heer, who edited this collection, and to Dan Wells and his gifted team at Biblioasis, who commissioned, designed and published it (and who saw the need to collect work from Canada's short literary non-fiction tradition in the first place).

The stories in this collection benefit from the wisdom of many brilliant editors and publishing professionals who didn't contribute directly to the pieces herein. My heartfelt thanks to Anne Collins, Craig Pyette, Lynn Cunningham, David Hayes, George Russell, Andrea Curtis and Dan Rubinstein.

As ever, my deepest debts of gratitude are to my family. My thanks to my parents, John and Margo Turner, who supported a fledgling freelance writer through many years of meagre and unstable income, and to my father-in-law, Bruce Bristowe, who has offered many kinds of support over the last ten years of a still wildly unstable freelance career in Calgary. Thanks to my aunts, Sharron Richards and Mary Hutchings, whose Toronto attic was the repository of those first revelatory issues of *Rolling Stone*. And my last and greatest thanks to my wife, Ashley Bristowe—my best editor, constant travelling companion, career manager and all-in-one support network—and my children, Sloane and Alexander, who are my greatest inspiration.

The articles in this collection originally appeared in the following periodicals:

"Flipflops, a Desktop and One Billion Reasons Never to Leave" originally appeared in *Shift Magazine*, issue 7.3 (May 1999)

"Take Me Down to Paradise City" originally appeared in *Shift Magazine*, issue 8.5 (June 2000)

"A Misunderstood Subculture, a Vegas Resort, and Lots of Black T-Shirts, Laptops and Booze" originally appeared in *Shift Magazine*, issue 9.5 (November 2001)

"Why Technology Is Failing Us (And How We Can Fix It)" originally appeared in *Shift Magazine*, issue 9.3 (September 2001)

"The Sgt. Pepper of Gaming" originally appeared in *Shift Magazine*, issue 7.2 (April 1999)

"Games Without Frontiers" originally appeared in *Shift Magazine*, issue 8.8 (October 2000)

"The Legend of Pepsi A.M." originally appeared in *This Magazine*, Vol. 36, No. 3 (November/December 2002)

"Buzz, Inc." originally appeared in *Shift Magazine*, issue 8.9 (November/December 2000)

"The Dirt on the Smoking Gun" originally appeared in *Shift Magazine*, issue 10.5 (December 2002)

"The Simpsons Generation" originally appeared in *Shift Magazine*, issue 10.3 (September/October 2002)

"The Age of Breathing Underwater" originally appeared in *The Walrus Magazine*, Vol. 6, No. 8 (October 2009)

"The New Grand Tour" originally appeared in *The Walrus Magazine*, Vol. 7, No. 4 (May 2010)

"On Tipping in Cuba" originally appeared in *The Walrus Magazine*, Vol. 9, No. 3 (April 2012)

"Calgary Reconsidered" originally appeared in *The Walrus Magazine*, Vol. 9, No. 5 (June 2012)

"Bearing Witness" originally appeared in *Eighteen Bridges* No. 3 (Winter 2011)